MUSHR

An Illustrated

FIELD GUIDE

BY NIKO SUMMERS

ILLUSTRATED BY JUNE LEE

Mushrooms: An Illustrated Field Guide

13-digit ISBN: 978-1-95151-131-9
10-digit ISBN: 1-951511-31-X

This book may be ordered by mail from the publisher. Please include $5.99 for postage and handling. Please support your local bookseller first!

Books published by Cider Mill Press Book Publishers are available at special discounts for bulk purchases in the United States by corporations, institutions, and other organizations. For more information, please contact the publisher.

Cider Mill Press Book Publishers
"Where good books are ready for press"
501 Nelson Place
Nashville, Tennessee 37214
Visit us online!
cidermillpress.com

Typography: Adobe Caslon, Caslon 540, DIN 2014,
Eveleth Clean Thin, Fontbox Boathouse Filled
Cover and interior design by Melissa Gerber

Printed in China
2 3 4 5 6 7 8 9 0

WARNING: Forage and eat wild mushrooms at your own risk!
Always check multiple guides, consult experts, and use the buddy system.

**"ALL FUNGI ARE EDIBLE.
SOME FUNGI ARE ONLY EDIBLE ONCE."**

—Terry Pratchett

CONTENTS

INTRODUCTION

Mushrooms are fun and exciting to find, and spotting them can be quite rewarding! Even for mushroom experts and enthusiasts, foraging can be tricky. But I believe that foraging is for everyone! It is an exciting way of looking at the world from a whole new perspective—a real bug's eye view. Mushrooms can be found in almost every climate in the world. So, they are easy to find no matter where you live. They are accessible even in the most unusual places. Mushrooms also have a rich history and have long been a part of our cultures and evolution as human beings.

Now more than ever, mushrooms and mushroom foraging have exploded in popularity, as the world has become desperate for a new perspective shift, and connection back to nature and the earth. This is what this field guide is for—an introduction to empower you to enter the wonderful and mysterious world of mushrooms and fungi, no matter what stage you are at. Within these pages you will find:

- Fifty beautiful illustrations of some of North America's most common and spectacular mushrooms
- Information to help you grasp the basics of identification
- Tips to help you find mushrooms
- Tidbits of mushroom science and conservation stories

Mushrooms: An Illustrated Guide aims to make essential North American mushroom information accessible and easily understood by anyone. We hope to spark your imagination and maybe help you discover your new favorite mushroom! Just one mushroom can start an entire obsession; for me, it was the maitake I found when traveling through Peru in the summer of 2019.

Have fun and explore the wonders and magic of nature with this book by your side!

—Niko Summers, Native Mushrooms, San Francisco, California

HOW TO USE THIS BOOK

When flipping through this book, you will find that the mushrooms are organized by alphabetical order. This will make it easy to read cover to cover or flip to a page and identify the mushroom you have foraged or found. Most mushrooms' names give a clue to their color; for example, yellow fairy cups, golden thread cordyceps, etc., are easy to find and remember. You can match the detailed illustration of each mushroom to identify one that you have found. We also include notes on lookalikes, so you can be aware of any false or similar mushrooms that may be poisonous or inedible.

Each mushroom entry details the habitat in which the mushroom is found, so you will have all the information you need to pick spots and times of the year to look for them. Mushroom hunting is done seasonally, so each entry notes which season of the year is best for which mushroom. The climate is also important to consider (although you will find many mushrooms throughout various climates); we have included notes on climate and rainy seasons when mushrooms tend to pop.

As you go out and use this book as your guide you will find a full description of each mushroom, its difficulty level, its typical size and dimensions, as well as a field notes section aiding you further in specific identification. The glossary in the back of the book defines key mushroom terminology.

THE PILLARS OF MUSHROOM IDENTIFICATION

The main things to look for when identifying mushrooms are:

STIPEND OR STEM: If they have one, what is the thickness and the shape? This will determine what kind of mushroom you are looking at and if it is a true or false (also known as a lookalike) mushroom.

GILLS OR A POROUS UNDERBELLY: When examining the underbelly of a mushroom it is important to note if it has true gills, small folds, or a porous underbelly (meaning it would be a polypore mushroom). Knowing this will further help you to identify what kind of mushroom you are looking at and develop a broader knowledge of mushrooms in general.

COLOR: When looking at any wild plants, we generally look at color as an indicator to determine what could be poisonous and what could be safe to eat. With mushrooms it is a little different, but not by far; for example, we all classically know amanitas for being bright red with white dots marking it as a poisonous or hallucinogenic species. That said, there are some mushrooms that have stunning colors marking them medicinal or edible as well.

CAP: The size, shape, and even texture of the cap is crucial to identifying mushrooms. This can determine many things. It can tell you what stage your mushroom is in; if your mushroom is edible, medicinal, or poisonous; as well as what kind of mushroom you are looking at. Observing the cap of the mushroom can even tell you the age, if it has rings spanning outward or elongating down.

CLIMATE: Another thing to be particularly aware of is where you are: is it a coastal region, is the climate very damp and wet, or are you in a more controlled microclimate? These things will determine where and what kind of mushrooms you may find in the area. Some mushrooms prefer more moist and damp climates where they are constantly getting a frequent supply of water through the air; others do not mind as much and like windier areas where there is more forest litter and mulch.

HOW TO START FINDING MUSHROOMS

Go for a walk, take a hike, or explore new places after a rainy season (or even in the rain!) and see what you find. Foraging in the rain can be a mystical and magical experience; I have found it to be surprisingly pleasant and meditative, and I have my own personal routine for rainy day forages.

It does not take much to find mushrooms if you are looking for them. I like to use the buddy system and always go with a mentor or mycologist if foraging for culinary or medicinal purposes.

Mushrooms love wind and rain. So, if you want a bounty, going after a rainy season or a few days thereafter is ideal. If you are nervous about foraging in the wild or further out of your vicinity, there are plenty of mushrooms that pop up in parks and other local places. Chicken of the woods is a common mushroom I find at many local parks, and I have cooked it several times safely.

RESOURCES

There is a plethora of resources online and various field guides that are great for learning more about mushrooms and identification. Here are some of my favorites:

BOOK OF MUSHROOMS: This is an amazing app for android or IOS, supplying the user with information on mushrooms they have found. Just use your camera to take a picture of the mushroom and it will tell you what it is! It is a handy tool for surprise findings.

FUNGI FOR THE PEOPLE: This grassroots organization encourages mushroom cultivation and education. They provide community outreach programs and are connected to other organizations that help veterans and disabled or mentally ill communities. They have also been focusing on BIPOC and POC communities and creating safe spaces and educational events throughout the Bay Area.

FUNGI PERFECTI (WWW.FUNGI.COM): This is Paul Stamets' brand of mushroom powders, tinctures, and foraging and cultivation supplies, which he started in 1980. It has since grown a considerable amount and is one of the top brands in mushroom science and testing. Stamets offers online resources like lectures, blogs, and write ups on new findings, and his discoveries with new and emerging mushrooms from all over the world.

THE MYCOLOGICAL SOCIETY OF SAN FRANCISCO (WWW.MSSF.ORG): This is a dedicated group of mycologists and citizen scientists that examine different species of mushrooms, save and work on genetics, as well as hold events for local foragers and mushroom enthusiasts in San Francisco and around the Bay Area.

THE NATIONAL AUDUBON SOCIETY: This is a trusted group that has various books dedicated to birding, and national field guides including *The Field Guide to Mushrooms.* This is one of my favorite foraging books and one I keep on hand regularly.

NORTH AMERICAN MYCOLOGICAL ASSOCIATION (WWW.NAMYCO.ORG): This association is a great organization based around the restoration of habitats, local and native fungi species, and promoting a healthy relationship to the soil and earth. They have conferences, local events, host forays, and more. They have various resources on their website for information, mushroom identification, and sustainability practices for mushroom foraging.

OUTGROW (WWW.OUT-GROW.COM): Outgrow is where I like to source my cultivation supplies. They have various resources, grow kits, tutorials, and videos.

SAFETY AND POISONOUS SPECIES

In each description of the mushrooms, we let you know which ones are poisonous, hallucinogenic (containing psilocybin, psilocin, or both), and simply inedible. Never forage for mushrooms and eat them alone, no matter how much experience you have. I always recommend using a buddy system or going with a mentor, and it is extremely important to do so before ingesting any wild plants, fungi, or mushrooms. *Never rely on one resource when foraging. You will want to consult multiple experts and refer to several trusted mycology books before even thinking about eating anything you find in the wild.*

AGARICUS

COMMON NAME
Agaricus

SCIENTIFIC NAME
Agaricus abruptibulbus

ORDER: Agaricales

FAMILY: Agaricaceae

Agaricus abruptibulbus is a commonly found North American mushroom. It is most typically known as the "flat-bulbed mushroom" or the "abruptly bulbous agaricus" due to its cap, which is either white or a brownish yellow. The mushroom grows up to four inches tall with a three-inch cap. When used for culinary purposes, it gives off a pleasantly light scent of anise.

HABITAT: Agaricus grows in dense forests that have many leaves on the forest floor, and especially around conifers. That is because conifers contain a thick, moist patch of ground cover underneath them. These mushrooms are found up and down the California coast and can also be found in Oregon, Washington, and Seattle.

DIFFICULTY LEVEL: Agaricus have been easily found from the California coast all the way up to New York, and throughout Mississippi, among other places. They are very easy to find when it comes to foraging. This specific agaricus, Agaricus abruptibulbus, can be found in places spanning from Mississippi to New York as well as Canada.

TYPICAL SIZE: Four inches tall with a three-inch cap.

NOTES: This mushroom was originally named by the mycologist Charles Horten Peck in 1900. It is known to absorb a toxin called cadmium, which naturally occurs in soil, quite fast. This means that the soil around this mushroom should have little to no cadmium. I commonly see these mushrooms around San Francisco parks, such as Golden Gate Park, where I have found them most frequently after the dew breaks and the rain showers have cleared. But I never consume them as they may contain cadmium.

LOOKALIKES: Agaricus silvicola or "wood mushroom" is a species of agaricus that is related to button mushrooms and grows in dense forests and leaf scattered moist areas. Commonly found in California, Oregon, and Washington. It can be easily confused for Agaricus abruptibulbus.

EDIBILITY AND USES: This mushroom is edible and has a slight subtle smell of anise when cut; can be used for culinary purposes.

WARNING: ***High risk of ingesting cadmium if foraged in littered forests, camping sites, cannabis grow zones, etc. Cadmium is found in batteries, pigments, and electronic shells or coatings; it gives them a metallic shine. Cadmium waste is typically either dumped or removed through a hazard waste facility, but pollution in deserted areas is common. Always forage with a buddy who is an experienced forager or buy from a farmers market where identification is guaranteed.***

FUN FACT: Once fully grown, Agaricus silvicola looks almost identical to Agaricus abruptibulbus. They are both found in the same family. Silvicola's cap folds inward and is slightly longer. Agaricus abruptibulbus develops yellowing on its cap with age, whereas the Agaricus silvicola stays white.

AGARIKON

COMMON NAME
Agarikon

SCIENTIFIC NAME
Laricifomes officinalis

ORDER: Polyporales

FAMILY: Fomitopsidaceae

DESCRIPTION: Agarikon is known for its beehive-like shape and conk structure. It is long and wide and it thrives in old growth forests. Traditionally, it has been found close to villages, in indigenous burial grounds, and forests. It has been used to treat against Influenza A and B, as well as HSV-1, HSV-2, and mycobacterium.

HABITAT: Agarikon is known to be found growing in old growth forests as well as on birch, redwood, and spruce trees. It tends to grow best in the pacific northwest, namely from Michigan to Washington, and then southward to California and Colorado. This mushroom is classified as a wood rotting fungus and grows on numerous varieties of wood, including Douglas fir, larch trees, hemlock, spruce, and conifers.

DIFFICULTY LEVEL: This mushroom can be tricky and rare to find. It can be a nice surprise when wandering the forests in Shasta County, California, or on the Oregon coast.

TYPICAL SIZE: Agarikon mushrooms can become very large, even as big as a beehive. In fact, they can often be mistaken for one depending on age, color, and size. This mushroom can become over twenty inches (fifty centimeters) long!

NOTES: Agarikon mushrooms are famous for their antiviral and antimicrobial properties. The human body needs to constantly have a strong immune system to respond to whatever gets thrown at it. With lifestyle changes, stress, diet, and everything else we take on as human beings, we can always gain some help from our plant and fungi allies. Agarikon is one of those allies that strengthens and tones the immune system. This fungus can live for over seventy-five years, and it has the ability to ward off natural pests and parasites that try to attack it. This makes agarikon the number one immune boosting mushroom in this book!

CONSERVATION STATUS: Endangered. (Please do not harvest or forage this mushroom.)

LOOKALIKES: There are no known lookalikes of agarikon to date.

EDIBILITY AND USES: This mushroom is known to be edible but not in large amounts, and is most commonly consumed in powders or capsules.

FUN FACT: Mycologist Paul Stamets once went on an expedition along the western coast of Canada in search of the agarikon mushroom. He hunted this mushroom for miles and miles, using a small sail boat to hug the coast, until he and his crew finally found it high up on a bluff. The specific species of agarikon he found only exists in the wild. Unfortunately, the sample did not survive.

AMETHYST DECEIVER

COMMON NAME	SCIENTIFIC NAME
Amethyst Deceiver	*Laccaria amethystina*

ORDER: Agaricales

FAMILY: Hydnangiaceae

Amethyst deceiver is a beautiful royal purple mushroom that lives up to its name. When spotted in the underbrush, it is a deep purple. As it dries out, the color fades. The gills are spaced further apart than most other mushrooms and are powdered with fine white spores. It is not typically a culinary mushroom because of its arsenic absorbing properties and tough exterior, which does not soften even after being cooked.

HABITAT: This mushroom can be found in coniferous forests as well as deciduous ones. It grows in a scattered pattern rather than in groups, and prefers oak wood. You might find it across the Great Plains, typically east where there are woodlands. It is more of an acid soil loving mushroom, and absorbs heavy metals like arsenic.

DIFFICULTY LEVEL: When amethyst deceiver grows on moss

in the spring months, it stands out, making it very easy to identify. In the drier seasons, the color dilutes from the center of the cap. The cap will pale to a full white once it is dried out, making it more difficult to find and identify.

TYPICAL SIZE: The size of the cap is up to two inches wide and the stem is around two to three inches long, and 1/4-inch thick.

NOTES: Amethyst deceivers are known for their arsenic absorbing properties, therefore can be harmful upon eating. Although a beautiful find, it should not be consumed because of the arsenic absorbing properties.

CONSERVATION STATUS: None.

LOOKALIKES: Mycena pura is another purple colored mushroom that has pale gills and a soft, less woody exterior.

EDIBILITY AND USES: This species is edible but is not a common choice based on its arsenic absorbing qualities. It has no known medicinal or therapeutic uses. Depending on where they are found, amethyst deceivers can be somewhat—if not very—poisonous to consume if the soil quality has or carries levels of arsenic.

FUN FACT: Amethyst deceiver loves soil, unlike many mushrooms, which typically live on the actual wood of trees. This mushroom has a special relationship with a forest's underground tree root system and will emerge from the soil above tree roots.

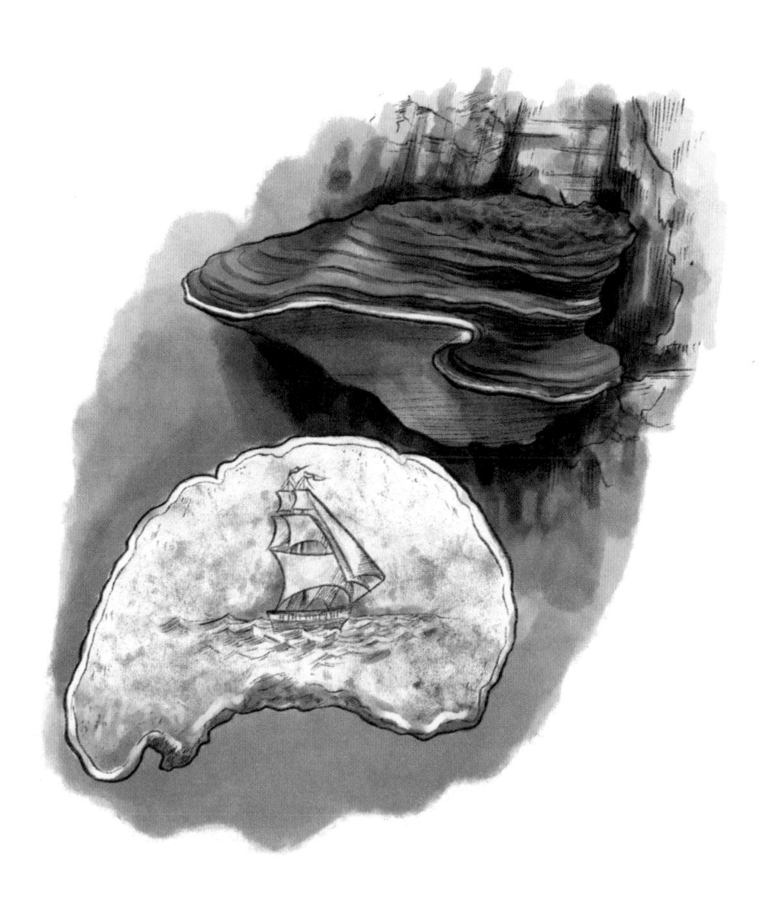

ARTIST'S CONK

COMMON NAME	SCIENTIFIC NAME
Artist's Conk	*Ganoderma applanatum*

ORDER: Aphyllophorales

FAMILY: Ganodermataceae

Artist's conks grow in the stumps of decaying or dead trees. They have a porous bottom that is easily scarred when drawn on with a stick or even your fingernail. They are called artist's conks for this exact reason. Whale artists would practice on them for scrimshawing whale bones, and some cultures would use them to practice on for tribal tattoo designs. They are rich in health benefits similar to reishi, and can take on the appearance of the stump they're growing on, making them a tricky find for beginners.

HABITAT: Typically, this mushroom grows in old conifer forests and on dead living stumps and trees. A good place to find this mushroom is in the Santa Cruz mountains among fallen old-growth redwoods, and across North America.

DIFFICULTY LEVEL: This is a more common species of mushroom, found in abundance throughout northern California.

TYPICAL SIZE: Ganoderma applanatum can grow up to twenty inches in length and eleven inches wide.

NOTES: Artist's conks are extremely medicinal. They have a plethora of benefits due to their compounds, which consist of triterpenes assisting in digestion, as well as beta glucans which help boost the immune system. This mushroom can have a strong laxative and diuretic effect. It is also anti-inflammatory, antibacterial, and known to be good for the lungs and upper respiratory system. It has anti-tumor and anti-fibrotic benefits, but more research needs to be done to confirm that.

CONSERVATION STATUS: None.

LOOKALIKES: There are a few polyporus varieties that can be hard to tell apart. The birch polypore and reishi can be hard to distinguish to the untrained eye.

EDIBILITY AND USES: This mushroom is typically boiled and steeped in a tea, although homeopathic or medicinal tinctures can be made as well.

FUN FACT: Artist's conks are a delightful find because their underbelly can be a marvelous natural canvas. It can be drawn, carved, and etched onto, and there is a long folk tradition of using it to make beautiful illustrations.

This mushroom is also purported to help assist cancer treatments. It makes a compound that may slow or stop the enlargement of polyps and cysts and reduce inflammation in the body.

BEEFSTEAK POLYPORE

COMMON NAME	SCIENTIFIC NAME
Beefsteak Polypore	*Fistulina hepatica*

ORDER: Polyporales

FAMILY: Fistulinaceae

Fistulina hepatica, also known as the beefsteak polypore, lives up to its name in that when sliced and eaten, this mushroom takes on the appearance and taste of a beef steak. It is spongy and porous inside, and it leaks out a crimson red liquid when cut. This species of polypore is commonly seen and foraged, although it can be difficult to prepare, requiring long cook or boil times before becoming edible.

HABITAT: This mushroom is found in the early months of August until late autumn. It can be found among oak and sweet chestnut trees. It is commonly found on the eastern side of the United States, from Florida up to Maine.

DIFFICULTY LEVEL: This polypore is a common find on living or dead wood on trees like oak, sweet chestnut, and on knots of Eucalyptus trees.

TYPICAL SIZE: The beefsteak polypore gets up to two inches thick and the cap up to eleven inches wide.

NOTES: Beefsteak polypores have a similar texture to meat. They have been used for meat substitutes in the past, and are used presently in modern gourmet cooking. When foraging and collecting this mushroom, you will want to grab it when it is still young, soft, and fleshy. This will make it easier to consume with a less bitter taste and cooking time. It does carry a subtly acidic and sour taste, making seasoning and flavor rubs a great addition.

CONSERVATION STATUS: None.

LOOKALIKES: Chicken of the woods (Laetiporus sulphureus), which is also an edible mushroom, can be a common lookalike for this species. Take care to notice detailed differences in shape and color between the two.

EDIBILITY AND USES: This mushroom is known for its culinary purposes. It looks like a slice of raw meat and is commonly used as a nutritious meat replacement. When cooking, use young (softer) polypores rather than the older (tougher) ones. This will result in shorter cooking time and better taste.

FUN FACT: When sliced and eaten, the beefsteak polypore takes on the appearance and taste of a (you guessed it) beef steak. It is spongy and porous and, when cut, leaks out a crimson red liquid.

BIRCH POLYPORE

COMMON NAME
Birch Polypore

SCIENTIFIC NAME
Fomitopsis betulina

ORDER: Polyporales

FAMILY: Fomitopsidacea

This polypore is a bracket fungus also known as birch bracket or razor strop, and grows preferably on birch trees. This fungi is a polypore showing a tinted white base or underbelly and a brown to fading gray cap. The birch polypore has tons of tiny tubes that each contain hundreds of its spores. This specific polypore isn't tough and woody like most, but rather soft and fleshy or sponge-like in texture. With age, this polypore can become tougher and rubberier after a year.

HABITAT: Fomitopsis betulina grows only on birch trees across Canada and the United States. They can be found in places that have had environmental stress such as fire, drought, or major suppression by deforestation and other trees in the area.

DIFFICULTY LEVEL: When foraging, these mushrooms can only be found on birch trees. Depending on the location, it can be a rare or more common find.

TYPICAL SIZE: The average size of this polypore ranges from four to eight inches across the cap and three to four inches in length.

NOTES: This polypore has an amazing number of medicinal benefits that can be widely used by herbalists, foragers, and survivalists. For one, the birch polypore carries a load of anti-bacterial, anti-microbial, and antiviral properties—making it the perfect aid in an apocalypse or just rummaging around in the woods.

CONSERVATION STATUS: None.

LOOKALIKES: None.

EDIBILITY AND USES: This polypore is inedible but is used for its medicinal properties. It can be used as a tincture to treat stomach parasites; made into a tea from dried material to be used as a laxative or to calm the nervous system. It can also be used as a bandage by slicing it into thin strips, drying it, and then applying it to a wound to stop the bleeding and protect against bacterial infections.

FUN FACT: Birch polypores have long been used medicinally in many cultures around the world because of their broad range of healing properties.

It is possible that it may have some psychoactive effects and has been used in shamanic contexts. Because it can grow to be extremely tough as it ages, it is sometimes known as razor strop mushroom, and used to sharpen blades. Thin slices of birch polypore could be used to stop a bleeding cut, and bigger pieces were often carried to use as tinder, as they make a handy fire starter. They are also thought to be antiviral and anti-tumor.

Although the birch polypore has been widely known and used for thousands of years, it caught international interest when it was found on the necklace of a centuries-old mummy discovered in the Italian Alps. Researchers suspect that Ötzi (as the mummy is called) may have used the mushroom medicinally, or as a blade sharpener, or both.

BLACK TRUMPET

COMMON NAME
Black Trumpet

SCIENTIFIC NAME
Craterellus cornucopioides

ORDER: Cantharellales

FAMILY: Cantharellaceae

The black trumpet mushroom, also known as trumpet of the dead or black chantarelle, is an edible mushroom that is similar to an oyster mushroom in taste and texture. It is found scattered on forest floors under broad leafed trees, like beech and oak. It is found in damper zones with moss and wet calcium rich soil. It is solid black and shaped like a small trumpet with a divot in the middle and a fanning cap going into a conjoined tubular stem.

HABITAT: This mushroom is found throughout North America and grows on the ground. It hides underneath leaf litter of wide branching oak trees and beech trees that supply it with a large canopy for rich, moist ground.

DIFFICULTY LEVEL: This mushroom can be found commonly, but is rarely sold at markets or through farmers for culinary use. Foraging has become more popular over the years for this mushroom, making spots to find it more accessible and used more often.

TYPICAL SIZE: These mushrooms are average sized and range from 2 to 4 inches tall and about ¼ to 2¾ inches across its fanning cap.

NOTES: This mushroom has many culinary uses. When dried out, it can have hints of black summer truffles, making it a great choice for umami dishes and other spice blends. A Portuguese study showed that 100 grams of these dried mushrooms contains a significant amount of protein (69.45 grams), which can replace meat for those wanting to lower their consumption while maintaining a high protein diet.

CONSERVATION STATUS: None as of yet.

LOOKALIKES: Black trumpet mushrooms can be commonly mistaken for Urnula craterium or the devils urn. Although similar in appearance, black trumpet has more of a cup-like display, and, while very bitter, it is not poisonous.

EDIBILITY AND USES: This mushroom is a great addition to soups, salads, and anything with umami spices and flavorings. It can also be eaten fresh and cooked with oil or butter to bring out its taste and flavor. When dried it can have subtle hints of black summer truffle, which can be crumbled on top of condiments or put into spices and flavorings. When cooked fresh it has an earthy, rich taste with the texture of chicken or lean meat. Sautée with garlic, onion, and tomato for a savory meal over rice or put it on a pizza!

FUN FACT: Black trumpets have an otherworldly beauty, but they can be tricky to find because of their dark color. As they mature, their shape takes on that of a flower in a gorgeous bouquet.

BLUING PSILOCYBE

COMMON NAME	SCIENTIFIC NAME
Bluing Psilocybe	*Psilocybe cyanescens*

ORDER: Agaricales

FAMILY: Hymenogastraceae

Bluing psilocybe, otherwise known by the names of psilocybe cyanescens, wavy caps, or potent psilocybe, can take on a bluing of the stem when picked, marking it a potent entheogen. This psychoactive mushroom can be found to be plentiful, with over three hundred mushrooms growing from the same patch at times. Although rather small in size, it can be edible when its psychoactive compound is parboiled out in hot water, but it is not commonly consumed.

HABITAT: These mushrooms can be found growing in bountiful amounts among woodchips and are known to be wood lovers. They can be found from British Columbia to San Francisco, popping up around the sides of garden beds where fresh mulch has been laid and a few seasons have passed.

DIFFICULTY LEVEL: They can be a common find in parks that have untreated mulch beds; neighborhoods with lots of landscaping and gardeners; as well as in dense forests with lots of bark that has fallen to create woody mulch patches.

TYPICAL SIZE: When still young, the bluing psilocybe is about half an inch in length; in full maturity, it reaches up to two inches across. The stem is two to three inches long and the cap will become wavy after fully breaking its veil.

NOTES: While technically edible, this mushroom has an extremely bitter taste and is not a good choice for culinary purposes. The psychoactive compounds in this species cause a blue stain when bruised or cut into, which is how to identify some psychoactive and poisonous species of fungi.

CONSERVATION STATUS: None.

LOOKALIKES: These mushrooms have many common lookalikes because their basic and common characteristics are similar to other species of fungi. One such lookalike is the Psilocybe allenii, which is commonly found in California and up to Washington.

EDIBILITY AND USES: This mushroom is edible but contains high amounts of psilocybin and is highly psychoactive. The mushroom may be parboiled in hot water to eliminate or reduce the hallucinogenic compounds if the consumer does not want a psychoactive effect. Otherwise, this mushroom is very bitter but can be used in psychedelic therapy and is easily cultivated in legal states.

FUN FACT: Bluing psilocybe is a heavy wood loving mushroom. Its preferred growing medium is wood chips, making it an easy find in gardens and local parks. Throughout the mycology community and beyond, bluing psilocybe is known for its psychedelic compounds. It carries a moderate level of psilocybin, making it a hallucinogen. This is a therapeutic choice among practitioners when working with clients in decriminalized and legalized states.

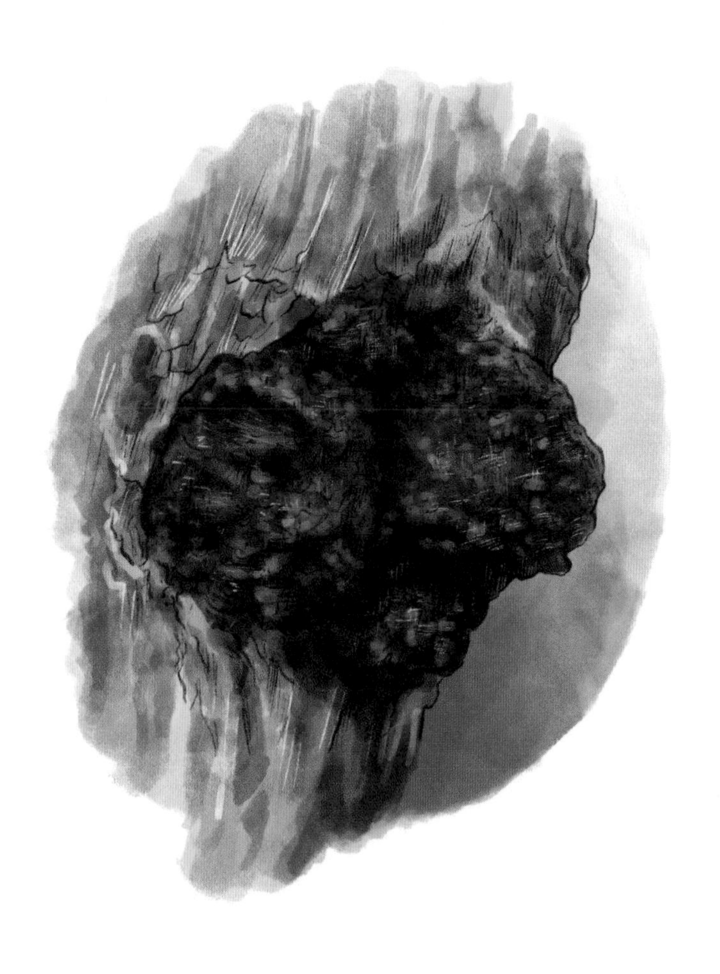

CHAGA

COMMON NAME
Chaga

SCIENTIFIC NAME
Inonotus obliquus

ORDER: Hymenochaetales

FAMILY: Hymenochaetaceae

Chaga is a fungus that is parasitic to birch trees, other similar trees, and trees growing in the same area. This mushroom is a conk species and has the characteristic of burnt wood. It is usually seen bulging out of a split branch or knot in the tree, possibly where wounds have been made and the fungus has a chance to infect its host.

HABITAT: Commonly found growing high up on birch, eucalyptus, conifers and other trees. Chaga can be found growing in redwood forests along the California and western United States coast all the way up into Canada and Alaska.

DIFFICULTY LEVEL: Chaga can be a rare find, making it expensive to buy and a sought after treasure when foraging. I still have yet to find this majestic beast.

TYPICAL SIZE: Chaga can be pretty big, averaging eight to twelve inches in width, and weighing up to five pounds.

NOTES: Chaga is becoming increasingly endangered due to its medicinal benefits, and because its taste is similar to that of coffee, making it a popular coffee alternative. Sustainability practices are being widely used and recommended when harvesting and foraging for this mushroom. Check to make sure that any chaga products you buy are produced sustainably.

CONSERVATION STATUS: Close to endangered.

LOOKALIKES: None.

EDIBILITY AND USES: Chaga is widely used for its medicinal benefits and for its taste as a coffee replacement. The tough exterior makes it difficult to cook, but if finely powdered, isolated in an extract, or made into a tea, it carries a wide range of anti-inflammatory, adaptogenic, and stress reducing benefits and qualities.

FUN FACT: Chaga is a world-renowned medicinal mushroom. It can assist the immune system with attacking various types of cancer cells and tumors, and it can help bring down overall inflammation in the body. Because of its bitter taste, it is usually drunk as a tea and mixed with honey or cocoa to make it more appetizing.

CHANTERELLE

COMMON NAME	SCIENTIFIC NAME
Chanterelle	*Cantharellus*

ORDER: Cantharellales

FAMILY: Cantharellaceae

The chanterelle mushroom is widely sought after and has been increasing in popularity over the years as foraging has become more popular. It can be identified by its folded gills that go into a forked shape at the end. The color is a golden yellow tinted body that can also appear more orange when cultivated.

HABITAT: Chanterelles are found across Canada and the United States. A few great places to look are throughout California, Oregon, Pennsylvania, and Washington. They like mossy damp areas that are frequently moist or have high levels of fog or moisture in the air.

DIFFICULTY LEVEL: These mushrooms are common to find throughout northern California and along the coast up to Washington.

TYPICAL SIZE: The fanning chanterelle cap can be as wide as 5 inches, 2 inches; is common to see in the wild, whereas 5 inches is more common in the cultivated variety.

NOTES: With a rich flavor and peppery notes, chanterelles are great mushrooms for culinary purposes. They are a top choice for chefs when the season comes around.

CONSERVATION STATUS: None.

LOOKALIKES: The false chanterelle is very similar in appearance and falls in the Cantharellaceae family, but is a beige to light tan color and has true gills as opposed to folds. Although edible as well, it may cause gastrointestinal issues. The jack-o'-lantern mushroom is another mushroom commonly mistaken for a true chanterelle, although it can be identified by its unforked true gills underneath. It is very poisonous but not lethal.

EDIBILITY AND USES: Chanterelles are commonly used for cooking and are used in a lot of gourmet dishes. They can carry a rich flavorful taste and can be prepared in many ways. The fruit body can be dried and added as a flavor enhancer for soup stocks, spices, and curries. The fresh mushroom can be sautéed, pan fried, or baked with oil or butter and salt to take out the water and enrich its flavor.

FUN FACT: Chanterelles can be dried and added as a flavor enhancer for soup stocks, spices, and curries. The fresh mushroom can be sautéed, pan fried, or baked with oil or butter and salt; this cooks out some of the mushroom's water and enriches its flavor. Chanterelles are an excellent addition to almost any dish!

CHICKEN OF THE WOODS

COMMON NAME	SCIENTIFIC NAME
Chicken of the Woods	*Laetiporus sulphureus*

ORDER: Polyporales

FAMILY: Fomitopsidaceae

Chicken of the woods mushroom, a.k.a. chicken mushroom, lives up to its name in that it tastes like chicken when cooked, and also resembles the poultry flesh when cut into fresh and after cooking. This mushroom is commonly found growing at the base of dead or decaying trees, flashing its bright orange and fading yellow to orange rings. It is most commonly found on eucalyptus and old redwoods or other trees. This mushroom has a porous polypore-like structure and can absorb oils from the trees it grows on. This makes the ones found on eucalyptus trees less than ideal for culinary purposes, as the inside will take on the taste of that tree.

HABITAT: Laetiporus sulphureus is found across North America throughout dying or decaying forests, as well as on stumps and logs that have been cut down due to logging or landscaping.

DIFFICULTY LEVEL: This mushroom is commonly found and is pretty easy to spot by its bright orange and yellow coloring.

TYPICAL SIZE: The chicken of the woods mushroom grows in shelves and is measured by the shelf; each shelf can average around 2 to 23 inches wide and up to 1½ inches thick.

NOTES: This mushroom is widely known for its culinary purposes. It is most commonly used as a meat replacement, and can take on the taste and texture of white meat (particularly chicken).

CONSERVATION STATUS: None.

LOOKALIKES: None.

EDIBILITY AND USES: This mushroom is widely used and known for its culinary benefits. It can be made into crispy "chick'n" nuggets, used as a meat replacement for soups, curries, and tacos, or simply tossed on a salad.

FUN FACT: I have found this mushroom a number of times and am always in awe of its beauty. Chicken of the woods is extremely absorbent, with the particular ability to absorb oils naturally produced by trees like the eucalyptus. This gives chicken of the woods some protection against predators so they don't get eaten while growing; however, the mushroom's oils can also be toxic to humans in large amounts and can turn the mushroom quite bitter.

COMMON INK CAP

COMMON NAME	SCIENTIFIC NAME
Common Ink Cap	*Coprinopsis atramentaria*

ORDER: Agaricales

FAMILY: Psathyrellaceae

The common ink cap or inky cap mushroom is known for the liquid-like ink coming out from under its gills. Inky caps are slender and long with an umbrella-like cap opening up to create a downpour of inky spores. The mushroom is small in size and has a gray and black striped appearance. This mushroom is also known to have an adverse effect when taken after consuming alcohol.

HABITAT: This mushroom can be found in areas that are heavily disrupted or have been hit by recent rainfall. They are seen growing in lawns that are heavily mowed, lots that are frequently driven on, and disturbed garden beds or parks throughout North America.

DIFFICULTY LEVEL: Very easy to find and identify.

TYPICAL SIZE: This mushroom is typically a 3.9 inch cap with a 2 to 6 inch stem.

NOTES: This mushroom is thought to help treat alcohol-related disorders. Inky caps produce a compound that can make the consumer sick when ingested along with alcohol. If no more alcohol is drunk the effects will subside within a few hours.

CONSERVATION STATUS: None.

LOOKALIKES: None. This mushroom exudes an inky liquid and actually digests its own cap after fully maturing, making it particularly unique. As it grows, the common ink cap, or inky cap, as it is sometimes known, has an elongated dark brown cap with white tufts of shag and a small thin stem. This makes it easily identifiable. It grows in clusters, dripping ink droplets from under its umbrella-like cap, which curls upward until dissolved in a pile of ink.

EDIBILITY AND USES: This mushroom is edible but not commonly used for culinary purposes.

WARNING: ***Do not consume common ink cap mushrooms along with any form of alcohol.***

FUN FACT: One nickname for the common ink cap is the tippler's bane. That's because it is poisonous when consumed with alcohol. Using wine in your recipe or having a beer with the meal can cause extreme gastrointestinal issues, at the very least. Perhaps that is why the common ink cap is not widely known as a choice edible, but it can be used as one if precautions are taken. The ink can be used in pasta dough, for example, and the mushroom itself works well in various dishes, especially slow cooked stews and curries. The common ink cap colors food and gives off an earthy taste.

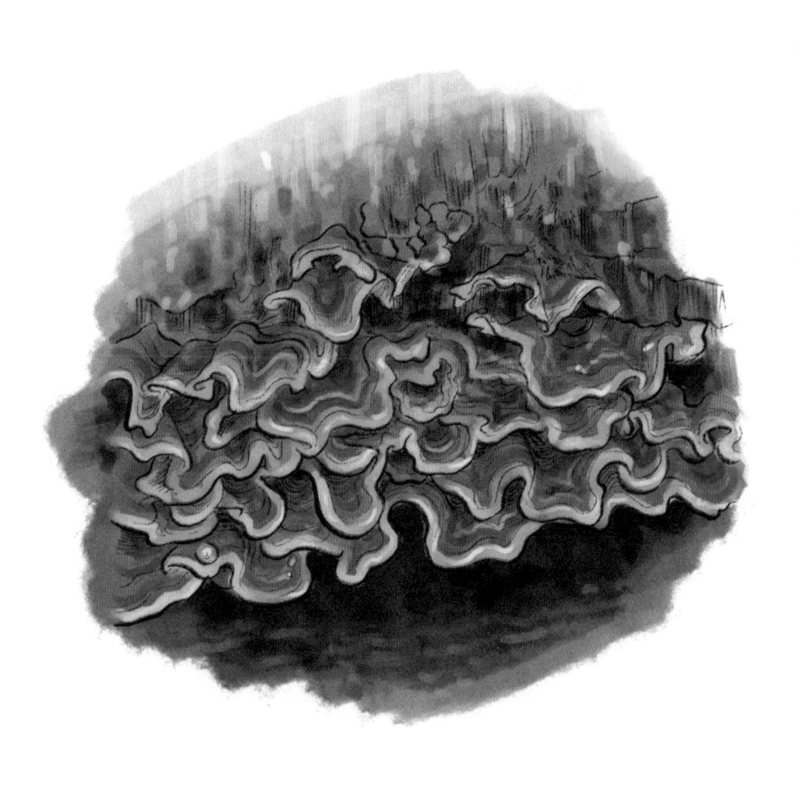

CROWDED PARCHMENT

COMMON NAME	SCIENTIFIC NAME
Crowded Parchment	*Stereum complicatum*

ORDER: Basidiomycota

FAMILY: Stereaceae

Stereum complicatum is a paper-like flakey crust fungus that likes to attach itself to the branches of peach trees and other wide-leafed trees. It gets its name from its similarity to crumpled parchment paper. When moist it can have a shiny and coral-like appearance. When dry it can be scraped off or cut. It is an inedible fungi that can infect and take over most trees on which it grows.

Crowded parchment fungi are found on stumps, scattered sticks along the forest floor, twigs, and branches of living trees, but they are particularly abundant on hardwood and oak trees.

HABITAT: Found in most areas commonly planted with trees, orchards, and even urban backyards all across the United States. They grow in abundance between New Hampshire and Massachusetts, southwest towards Georgia. You can even spot them in a few places in Mexico like Mexico City and Oaxaca.

DIFFICULTY LEVEL: Very common and easy to find on most trees (alive or dead).

TYPICAL SIZE: The shelf-like caps measure around 0.3 to 1.5 centimeters wide.

NOTES: Not much research has been done on this fungi and its uses.

CONSERVATION STATUS: None.

LOOKALIKES: Certain mosses, crust-like molds, and fungi, when grown in damp and rainy climates, may be mistaken for turkey tails or false turkey tails; crowded parchment is one of them. Luckily, it has a distinctive coloring pattern different than that of true turkey tails.

EDIBILITY AND USES: Inedible.

WARNING: ***Do not attempt to eat this fungi; insufficient research has been done on this species.***

FUN FACT: Crowded parchment is sometimes called a non-bleeder, meaning that it is quite dry. It is often brittle and crusty, or, in technical terms, resupinate. It typically forms a crusty layer all over a log and then fills in with more layers or shelf-like shapes.

CUCUMBER-SCENTED

COMMON NAME
Cucumber-Scented

SCIENTIFIC NAME
Macrocystidia cucumis

ORDER: Agaricales

FAMILY: Marasmiaceae

This mushroom's scientific name is Macrocystidia cucumis, and cucumis means "of cucumber" in Latin. The cucumber-scented mushroom is very small and stout when it is fully matured and fruited. It is an inedible mushroom although its name may indicate otherwise. It is not advised for cooking—no matter how good it smells! This mushroom displays pinkish brown spores when making a spore print and has a light cucumber scent that may verge on slightly fishy at times.

HABITAT: It can be found throughout North America, usually in parks and alongside or in garden beds. It is not as commonly seen throughout forests but can occur in the underbrush of pine needles and forest litter.

DIFFICULTY LEVEL: This mushroom is a common find among foragers and average joes who just like to garden. They can be easily found popping up in garden beds and in healthy rich soil.

TYPICAL SIZE: The cap of this mushroom can get up to 2 inches across in diameter. The stem is around 3 inches long at its full length and gets up to 5 millimeters thick.

NOTES: This mushroom generally grows alone or in small groups along trails. It tends to pop up around newly planted areas and is found in the warmer months of summer and fall.

CONSERVATION STATUS: None.

LOOKALIKES: There are a few lookalikes with some distinctive signs that show they are not the true cucumber-scented mushroom. One is the Flammulina velutipes, which grows on trees and has a white spore print, but does not smell of cucumber when picked. The other close relative is Pluteus cervinus, which grows on rotting trees and has a flatter shaped cap and lighter colored spores.

EDIBILITY AND USES: Although this mushroom may smell like something tasty, it is not recommended for consumption or culinary use.

WARNING: ***Cucumber-scented mushroom is classified as an inedible mushroom.***

FUN FACT: This mushroom is named after its scent, as it gives off a light cumber smell when fully matured. Although named after a cucumber, which is technically a fruit in the melon family, this mushroom is in fact not edible. Because of its small size, it would not yield a substantial portion anyway.

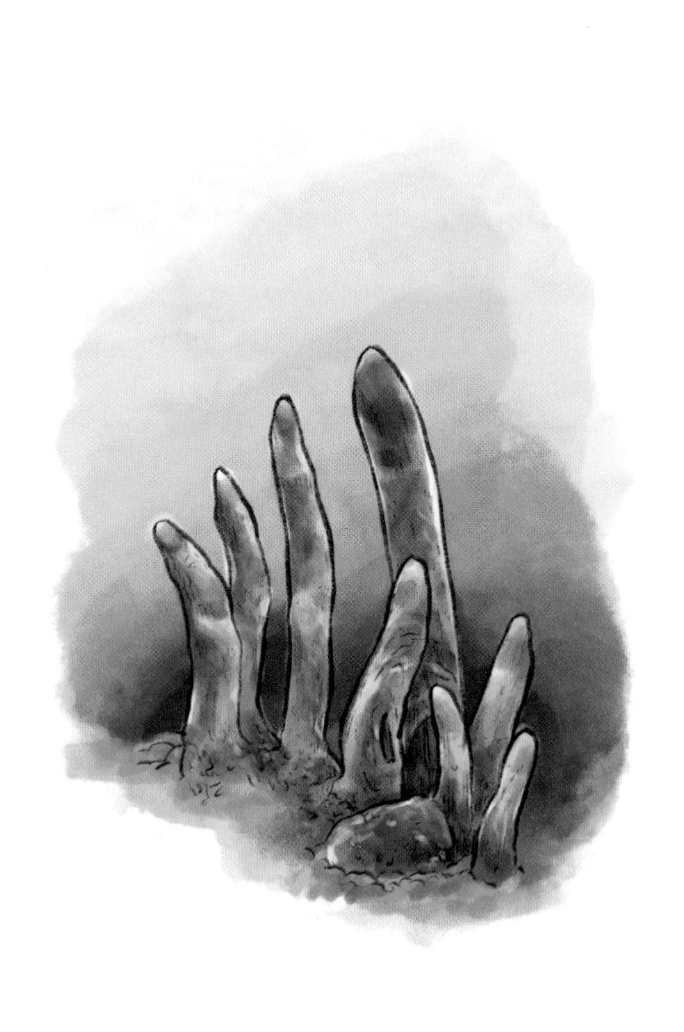

DEAD MAN'S FINGERS

COMMON NAME	SCIENTIFIC NAME
Dead Man's Fingers	*Xlaria polymorpha*

ORDER: Xylariales

FAMILY: Xylariaceae

Dead man's fingers is a very interesting mushroom. The name polymorpha means that this mushroom has various types of characteristics and traits as well as various species. This mushroom lives through its sexual organs which are long cylinder-like tubes that encase parts of decaying softwood trees. When producing spores this mushroom can take up to four months before the spores are ready to drop. The head or main structure of this mushroom is club-shaped with long rounded tips representing that of a corpse hand.

HABITAT: These mushrooms are usually found in densely populated forest floors and around decaying or dead softwood. They like fruit and maple trees. They can be spotted from Minnesota eastward to Maine, and along the southern coast of the United States, most commonly in summer and autumn.

DIFFICULTY LEVEL: This mushroom is commonly found throughout North America, but because of its color and size, it can be easy to miss or mistake for dead wood or decaying leaves.

TYPICAL SIZE: The finger-like structures on this mushroom are about 1½ to 4 inches tall.

NOTES: This mushroom has a very long life span; it can take up to four months to produce spores and can live ten years throughout the hyphae.

CONSERVATION STATUS: None.

LOOKALIKES: Watch for Xylaria longipes, or dead Moll's fingers, a slender and tall mushroom that mimics the club-shaped form of dead man's fingers. They often reside in the same areas, climate, and seasonal conditions, which can make them tricky to identify. But the dead man's fingers' coloring is rather distinct.

EDIBILITY AND USES: These mushrooms are inedible and can make the consumer sick if eaten. There have been a few reports of this mushroom being eaten raw—and actually enjoyed—using the young and tender fruits. But it is not recommended that you try it.

FUN FACT: The name polymorpha means "many shapes" and refers to the many unusual forms these mushrooms can take, not from their varied texture.

DESTROYING ANGEL

COMMON NAME	SCIENTIFIC NAME
Destroying Angel	*Amanita bisporigera*

ORDER: Agaricales

FAMILY: Amanitaceae

The destroying angel or Amanita bisporigera is (as you can tell from its name) in the amanita family. When growing and maturing into its fruiting body, this mushroom creates a white egg-like shell, which it then sheds. Bits of the shell can be found around it and on top of the cap as a helpful marker for identification.

Destroying angels may look like fairly edible and harmless button mushrooms or agaricus mushrooms; however, they are actually known to be one of the most fatally poisonous mushrooms in the world, causing the most human and animal deaths in fungi history. Once consumed, this mushroom starts creating severe and irreversible damage to the liver and kidneys. Symptoms are usually not present until twenty-four hours later. ***Beware of these extremely deadly and dangerous mushrooms.***

HABITAT: Destroying angels have long been found in various places all over the world. They are also known as eastern

Northern American destroying angels, because that is where you will find many of them: in the eastern United States. They are most commonly seen in dewy meadows, close to shrubs, and in less populated forested areas.

DIFFICULTY LEVEL: These mushrooms are very common to find and can be found even in lawns and local parks.

TYPICAL SIZE: The cap on this mushroom can be anywhere from 2 to 4½ inches wide, and the stem is 3 to 8 inches long.

NOTES: As mentioned above this mushroom is ***extremely poisonous*** and often deadly. That being said, there had been studies using specific compounds of a milk thistle extract, which was shown to be useful in patients who had been poisoned with the compound that is deadly in the amanita family called amatoxin; the treatment resulted in no deaths. So, although its medical powers merit further research, do not even think about picking or tasting this mushroom.

CONSERVATION STATUS: None.

LOOKALIKES: Since the Amanita bisporigera is an all-white mushroom it can have many lookalikes if no distinct details are noted. Meadow mushroom or Agaricus campestris is another common mushroom that can be found in similar habitats and is also related to the button mushroom.

EDIBILITY AND USES: Reader, beware! This mushroom is inedible; in fact, it is so poisonous to humans (and many other creatures) that it is one of the deadliest organisms on the planet. Even a small nibble of the cap can prove fatal. In humans and other mammals, the destroying angel's poisoning symptoms do not appear right away. They can take 5 to 24 hours to manifest, by which time the toxins are typically already absorbed, making the destruction of liver and kidney tissues irreversible.

WARNING: ***Amanita bisporigera is known as the destroying angel for a reason. If you see one, observe its deathly beauty from a distance. Do not touch or pick it.***

FUN FACT: As previously mentioned, this mushroom is extremely poisonous and often deadly. Destroying angels are tricky; they may look like edible and harmless button mushrooms or agaricus mushrooms, however they are actually one of the most fatally poisonous mushrooms in the world. They have caused by far the most human and animal deaths in fungi history.

DUNG-LOVING PSILOCYBE

COMMON NAME	SCIENTIFIC NAME
Dung-Loving Psilocybe	*Psilocybe semilanceta*

ORDER: Agaricales

FAMILY: Hymenogastraceae

Psilocybe semilanceata is one of the most commonly found psilocybin mushrooms known to date. They have a witch's hat-like appearance when fully matured, with a small nipple on the top and a long bell shaped cap. They can be found in wide cattle pastures in clumps of grass, sometimes hiding in fields and meadows near water-rich grasses and reservoirs. Dung-loving psilocybe are some of the most potent and active mushrooms containing psilocybin and baeocystin, a highly psychoactive compound in certain mushroom species. These mushrooms have an interesting symbiotic relationship with the roots of decaying plants, breaking down the natural matter. This makes them common to see in lawns and garden beds.

HABITAT: Most commonly found across North America in pastures that contain cows and sheep. They can also be found in meadows, on the edges of wetlands, and garden beds throughout the continent.

DIFFICULTY LEVEL: These mushrooms are very common to find, growing in many different climates all over the world.

TYPICAL SIZE: The cap of the dung-loving psilocybe ranges from 0.2 to 1 inch in diameter, and the stem can be 1 to 6 inches long and widens at the base.

NOTES: This is one of the most widely distributed mushrooms when it comes to foraging because of its potency. It has been used safely in small amounts for therapeutic use, but can also be very damaging to the vital organs if ingested in large quantities over a long period of time.

CONSERVATION STATUS: None.

LOOKALIKES: The common lookalikes for this mushroom are the Psilocybe pelliculosa and the Psilocybe mexicana. Both of these mushrooms are highly psychoactive and are not recommended for consumption or culinary purposes. The distinction between the Psilocybe pelliculosa and mexicana is their caps; the mexicana has a wider, UFO-shaped cap, while the pelliculosa has an elongated, bell-like cap.

EDIBILITY AND USES: This mushroom is highly psychoactive and is not recommended for use. Sale and distribution in many states is illegal, but in decriminalized and complying countries and states it can be used medicinally for therapeutic use.

FUN FACT: Entheogenic fungi are also known as therapeutic mushrooms. They carry a compound called psilocybin, which is the most active compound in magic mushrooms that causes the "tripping" effect. Magic mushrooms are known in almost every culture and religion throughout the world. They have a rich history in North America, specifically in San Francisco, California, in the 1960s and '70s hippy subculture. But they were not discovered by the hippy movement. In fact, they have been used by indigenous Americans for centuries. For example, entheogenic mushrooms were used in ancient cacao ceremonies throughout South America. They were buried with royalty and shamans. And they have long been a sacred tool and medicine to many cultures around the world.

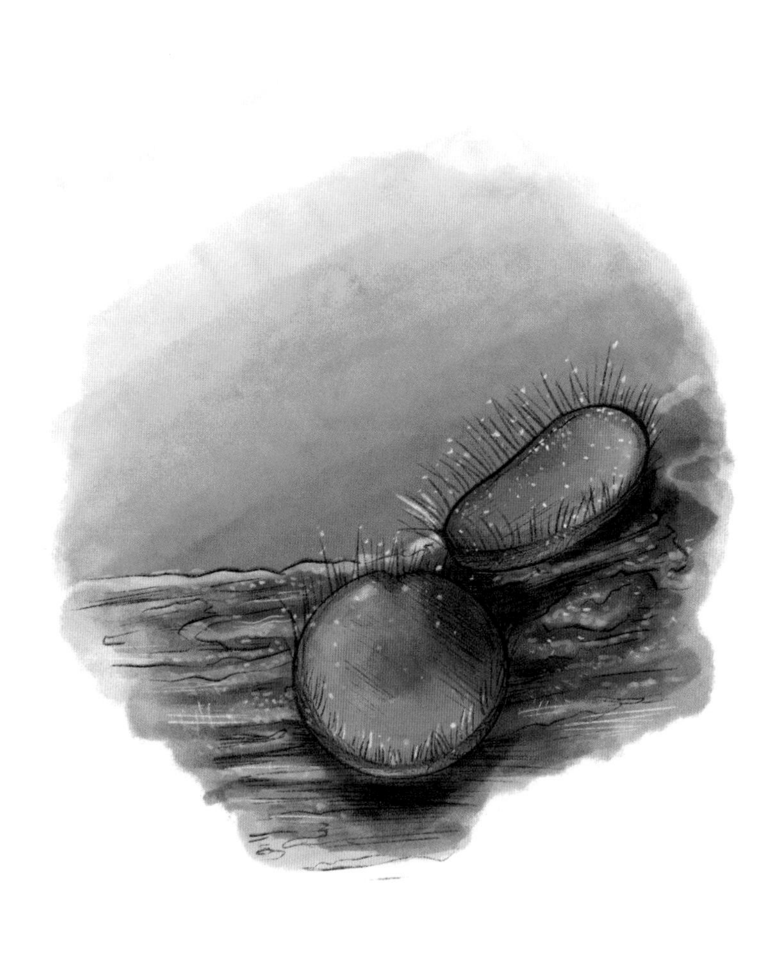

EYELASH CUP

COMMON NAME
Eyelash Cup

SCIENTIFIC NAME
Scutellinia scutenellata

ORDER: Pezizales

FAMILY: Pyronemataceae

The eyelash cup mushroom is a completely round and flat, or cup-like, shape. The inside of the mushroom is a bright orange-red where the spores are produced and released. The eyelash-like hairs around the edges of the mushroom are coarse and stiff, and the outside of the mushroom is a dark shade of brown. This mushroom can be found on fallen logs covered in moss on the forest floor, and normally grow in clusters. It is one of the prettier mushrooms to find, and is fairly easy to come across because of its bright coloring.

HABITAT: This mushroom is found in various parts of the world, but can be found in the winter and spring in North America. It is generally found in high altitude climates, growing on decaying fallen wood or wet leaf litter from forest floors.

DIFFICULTY LEVEL: These mushrooms are not difficult to

find but can be overlooked because of their small size.

TYPICAL SIZE: These mushrooms remain rather small, ranging anywhere from 0.2 to 1 centimeter in length, and up to 1 inch wide.

NOTES: This species of mushroom does not have a stem, making it one of the more interesting mushrooms I have come to enjoy and look for on forages.

CONSERVATION STATUS: None.

LOOKALIKES: Scutellinia umbrarum, Melastiza chateri, and Lamprospora are a few species that can look very similar to the eyelash cup, but differ in size. All of them scale much larger than the eyelash cup and have different coloring. They also have imperfections, including waves and deep creases in the caps.

EDIBILITY AND USES: This mushroom is too small to be edible and has no distinctive taste or smell.

FUN FACT: The forest is always watching! This pretty fungal gem has a memorable appearance with wispy eyelashes and a bright red cap. It is one of the more interesting mushrooms I always look for on my forages, as it is native to most coastal areas and high altitude climates. While not edible or medicinal, the eyelash cup mushroom is always beautiful to look at, and it can put you in awe of the wonder of nature.

FAIRY RING

COMMON NAME	SCIENTIFIC NAME
Fairy Ring	*Marasmius oreades*

ORDER: Agaricales

FAMILY: Marasmiaceae

Fairy ring mushrooms are a sight to see. They grow in a wide circle, almost as if by magic, and look like something straight out of a fairytale. These mushrooms have over sixty different kinds of species. One of the most commonly found and well-known is the scotch bonnet, an edible mushroom also known as fairy ring mushroom champignon or Marasmius oreades. Fairy rings can get very large—up to a few feet wide around bases or stumps of trees and in open grasslands that have been grazed on.

HABITAT: Fairy rings are commonly found in grasslands, meadows, fairgrounds, parks, and lawns; they have been found all throughout North America, coast to coast. They are also found in several other places throughout the world, including Ireland, Germany, the United Kingdom, Mexico, and Australia.

DIFFICULTY LEVEL: These mushrooms are very common to find and can be easy to spot because of their distinctive circular pattern.

TYPICAL SIZE: Fairy rings can grow up to 33 feet wide in diameter depending on the location, soil activity, and nitrogen in the soil.

NOTES: These mushrooms are near and dear to my heart. Ever since I was young I had a fascination and deep love for the mythical and symbiotic biological world we are all born into. I would sometimes stumble upon fairy rings in the park and would be intrigued by their majestic performance. They really are a spectacular sight to stumble upon, and are heavily spoken about in fairy magic, folklore, and mythology. Fairy ring mushrooms are intertwined in many beliefs on witchcraft, as well as the dancing of elves or fairies. Whatever you believe, these mushroom rings are beautifully magical and hold a high place in the spiritual and mythological world.

CONSERVATION STATUS: None.

LOOKALIKES: The Clitocybe dealbata is commonly mistaken for a fairy ring mushroom, and beware: It is poisonous, and it is also known as the sweating mushroom due to its toxic effect on the human body. It is a small white mushroom that grows near fairy rings and can look quite similar to them, so it is often picked with them by mistake. Here is one way to distinguish it from the fairy ring mushroom: when the Clitocybe dealbata is fully mature, its cap starts turning upward in a taco shape.

EDIBILITY AND USES: The most common and edible mushroom in fairy rings is the Scotch bonnet, which can be cooked but is not a widely known variety among many culinary artists.

FUN FACT: Fairy rings are an interesting phenomena. Because of their ecological impact and ancient folklore, they can evoke the splendor and curiosities of life itself.

In the Native American Ojibway folktale, *The Star Maiden*, a woman lived in a star that came down to an Ojibway village on Earth. She was in awe of the Ojibway village's peaceful lifestyle and special connection to the fairies. The fairy rings are mentioned in this story as a potential home for the Star Maiden.

Ecologically, fairy rings grow through a circular web of mycelium that is connected underground. After one year, even more mushrooms grow in the circular patterns, which can pop up like magic in parks and open fields.

FLY AGARIC

COMMON NAME	SCIENTIFIC NAME
Fly Agaric	*Amanita muscaria*

ORDER: Agaricales

FAMILY: Amanitaceae

This is another all-time favorite mushroom! It is otherwise known as the Santa Claus mushroom, or, scientifically, amanita muscaria. I love this ancient mushroom because of its bold color, its folklorish history, and its sacred ceremonial use. It is rich in history, magic, and splendor. The Amanita muscaria is a stunning red mushroom with specks of white atop its cap. There are a few subspecies of this amanita and they all result in different colorings.

HABITAT: This mushroom is usually found in woodland areas in pine needles and damp underbrush. In summer and fall, this mushroom can be found across North America, especially right after a good rainstorm. It is also commonly spotted in Asia and Europe, especially in Japan, Russia, Italy, Lithuania, and Finland. However, it has also been found in warmer climates, such as Mexico.

DIFFICULTY LEVEL: Because of its color and size this mushroom is a pretty easy find.

TYPICAL SIZE: The typical size of this mushroom's bright red cap measures from around 3 to 8 inches wide, and the white stem can get anywhere from 2 inches to almost 8 inches tall.

NOTES: This amanita is rarely fatal but can be quite dangerous and should not be picked or eaten. It is known to be hallucinogenic, and was used in Indigenous tribes and groups for various rituals and ceremonies. It is commonly found and gifted among mycologists and the entheogenic community as a token of good luck and a mushroom totem.

CONSERVATION STATUS: None.

LOOKALIKES: Many lookalikes have a similar appearance to this mushroom but usually have distinctively different coloring, from white to almost yellow. The panther mushroom, Amanita pantherina, is the most common and is also poisonous upon consumption.

EDIBILITY AND USES: It has a beige cap that can resemble a very faded red. It should not be picked or eaten. Although the Amanita muscaria is an occasionally deadly poisonous mushroom, it also has a history of being ingested and even enjoyed in many countries. It is typically eaten in very small amounts when the hallucinogenic compounds have been parboiled out.

If properly prepared and cooked, fly agaric can potentially adapt into a very potent medicinal mushroom. Some say it reduces inflammation, acts as a natural antibiotic, has anti-tumor benefits, and even helps with cognitive functioning.

A typical preparation involves taking the caps from the younger mushrooms and parboiling them at least two or three times, using lots of water, for two to four hours each time. It is also important to change the water and rinse the mushrooms in between each boil, as this can help remove most of the psychoactive properties. Some people then use the prepared mushrooms in very small amounts, by frying them up, preserving them as pickles, or even dehydrating them to make a tea. Regions in Italy and Russia are famous for doing this with fly agaric mushrooms in particular. It is not so widely known in North America.

WARNING: ***Although this mushroom is often portrayed as more psychedelic than deadly, it is important to note that this species is extremely poisonous and should be avoided. Because it is quite dangerous, this mushroom is not widely used in psychedelic therapy or cultivated.***

FUN FACT: Fly agaric is commonly known throughout pop culture as the psychedelic mushroom. It is also the classic mushroom depicted in the Mario series of video games. It can be seen in various classic paintings and even religious murals and artistry.

GIANT PUFFBALL

COMMON NAME	SCIENTIFIC NAME
Giant Puffball	*Calvatia gigantea*

ORDER: Agaricales

FAMILY: Agaricaceae

Giant puffballs are round or egg-like shaped mushrooms with no stem, usually protruding from a forest floor. They can be a fascinating mushroom when found fully mature as they can reach up to almost 3 feet in diameter and can weigh up to 44 pounds! After reaching full maturity, the mushroom will start to decompose, leaving the inside a greenish-gray. At this stage the mushroom is inedible and unsafe to eat. The younger puffballs are white inside, resulting in an edible mushroom that is safe to eat and widely foraged for this culinary aspect.

HABITAT: This mushroom is commonly found on forest floors that are densely populated, in fields or open meadows.

DIFFICULTY LEVEL: This mushroom is commonly found throughout North America and Europe.

TYPICAL SIZE: Ranging from 1 to 3 feet wide in diameter.

NOTES: This mushroom was known to be harvested and prepared (cut in thin slices) for medicinal use on battle wounds. It can be prepared by slicing the mushroom when young and white on the inside in 3 centimeter thick strips that can be applied underneath a wrap or bandage.

CONSERVATION STATUS: None.

LOOKALIKES: The giant puffball can look similar to the earthball Scleroderma citrinum; this mushroom is poisonous and can be intoxicating when consumed. The inside of the earthball Scleroderma citrinum is dark purple to black in its earlier stages of growth.

EDIBILITY AND USES: Giant puffball mushrooms are edible when young and still white on the inside. They can also be used to help dress wounds at this stage of maturity.

FUN FACT: Making more giant puffballs is delightful! As Dave Taft wrote in *The New York Times*, "Rain drops release many spores, but the flick of a finger can create a sinuous puff of dark brown smoke, potentially wafting the fungal spores into the lightest breeze and spreading puffballs near and far."

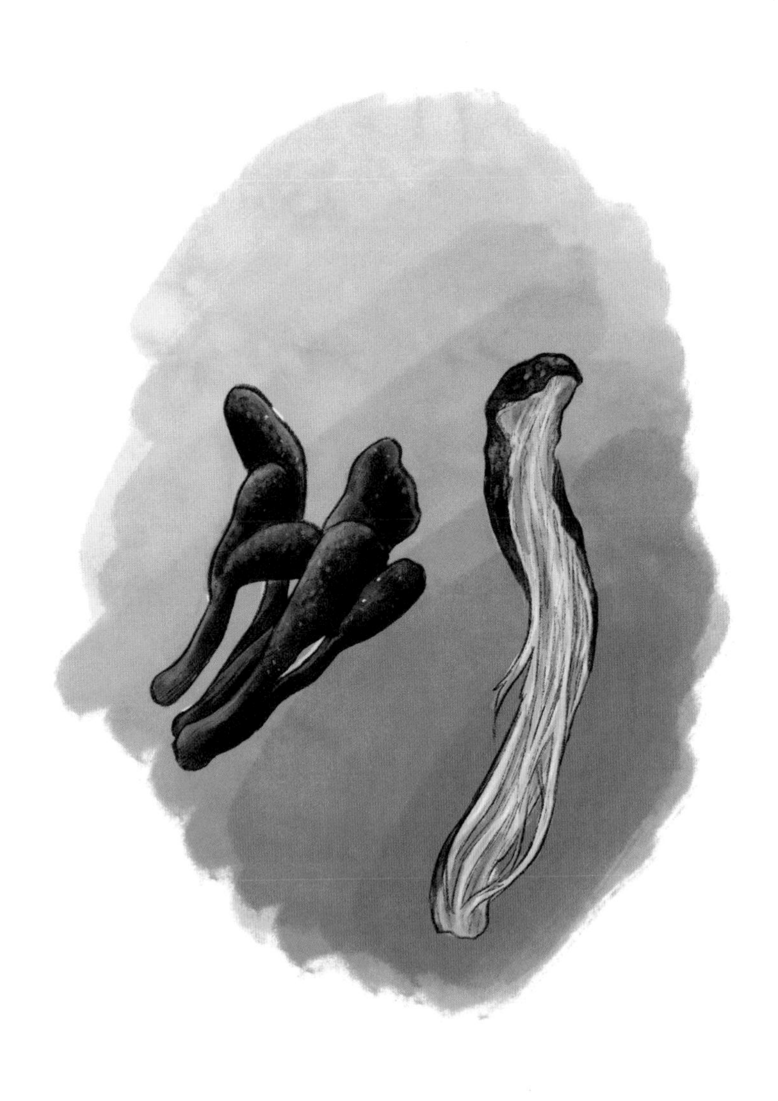

GOLDEN THREAD CORDYCEPS

COMMON NAME
Golden Thread Cordyceps

SCIENTIFIC NAME
Tolypocladium ophioglossoides

ORDER: Hypocreales

FAMILY: Ophiocordycipitaceae

The golden thread cordycep is a fungus that attaches an orange-gold thread of mycelium to its host that it then uses to grow and fruit out of. Unlike mycelium tied to rooting structures, decaying wood, dung, or soil, this fungi has learned and adapted to live off of an insect host. This mushroom grows as an independent tubular-shaped fungi and is always attached to an insect or fungal host. When foraging for this mushroom it is always fun to see what it is attached to in order to better identify the mushroom. It is pretty cool to see what you find! To do this you will want to: follow the cord of mycelium gently, remove the soil around the cordycep, and lightly dig or brush away the dirt. You should find a decaying insect or fungal host attached to the mycelium that you can collect for further inspection.

HABITAT: Golden thread cordyceps are commonly found all over North America and also in parts of Asia.

DIFFICULTY LEVEL: Golden thread cordyceps are often difficult to find because of their delicate golden color and attachment to a host.

TYPICAL SIZE: The small fungi range from 1 centimeter to 5 centimeters long.

NOTES: These cordyceps are believed to have various medicinal properties, but more research needs to be done. Dating back to ancient times many royalty in China would take cordyceps to help increase longevity; Chinese athletes were also known to take it in order to increase stamina and endurance.

LOOKALIKES: The velvety earth tongue and dead man's fingers are commonly mistaken for these; both are inedible and can, in fact, be quite poisonous.

EDIBILITY AND USES: Although they may be used medicinally, these mushrooms are not edible and can have a hard and tough interior. In Asia, some believe that golden thread coryceps strengthen the immune system, have anti-aging benefits, provide increased energy and endurance, and may help repair damage to the lungs and liver. Golden thread cordyceps are consumed as a health food in China, Korea, and Japan. They are sold as capsules and tonics to be used in traditional Chinese medicine (TCM).

FUN FACT: The golden thread cordyceps is an interesting fungi to examine. Most cordyceps attach onto a host plant, taking over its body and slowly shutting it down. The golden thread cordyceps expels an orange-gold thread of mycelium into its host that it then uses to grow and fruit out of.

GOLDEN WAXY CAP

COMMON NAME	SCIENTIFIC NAME
Golden Waxy Cap	*Hygrocybe chlorophana*

ORDER: Agaricales

FAMILY: Hygrophoraceae

The golden waxy cap mushroom is a beautiful, stunning golden orange mushroom, which is protected in parts of the world and is on the "red list" for threatened species in Germany, Switzerland, and Poland. Its cap, stem, and gills are all the same golden tone, and this mushroom is fully coated in a wet, wax-like film; this means that it is continually quite moist and often has a slippery or wet cap—hence the name.

Waxy cap mushrooms are known to arrive in autumn and early winter. Supposedly a cold shock hits and wakes up dormant mycelium underground, forcing it to start fruiting.

HABITAT: These mushrooms are most commonly found in wooded areas and dense forests throughout North America.

DIFFICULTY LEVEL: These mushrooms can be more difficult to find because of their stout, short size and appearance.

TYPICAL SIZE: The cap of the golden waxy cap mushroom ranges from 2 centimeters to 4 centimeters in diameter, and the stem is from 2 inches to 4 inches tall.

NOTES: Looking at the golden waxy cap's gills is a great way to distinguish from lookalikes; plus, they are very interesting to look at! They are slightly separated at the creases and can contain a different pattern than the typical mushroom gill structure. The gills on this mushroom tend to move down toward the stem, much like the unusual gills of oyster mushrooms.

CONSERVATION STATUS: This mushroom is on the red list of threatened species in Germany, Switzerland, and Poland. Unfortunately, there is no known conservation status in the United States.

LOOKALIKES: Hygrocybe glutinipes has a similar form and shape, but is a bit smaller in size, with a more glutinous cap and stem. The hygrocybe ceracea is another similar species but differs in that it is drier and typically smaller as well. Plus, there are a handful of lookalikes that are all ruddier than the golden waxy cap. Hygrocybe splendidissima is a small orange stalked mushroom with an orange-red cap and more widely spaced gills underneath. Hygocybe punicea is a shorter mushroom with a wide, reddish cap. Hygrocybe coccinea, also known as scarlet wax cap and scarlet hood, has a fully red stem and cap with a wide gill set. Hygrocybe conica, also known as witches hat or blackening wax cap, is a short and small orange mushroom with a peak on the top of its cap like a witch's hat; it has a bright orange appearance that can vary with the cap turning redder as it reaches maturity and blackening if bruised. These lookalikes can be hard to tell apart because of their orange and red color.

EDIBILITY AND USES: This mushroom can be foraged and eaten, but the tough exterior and slimy texture make it unworthy of cooking and it is not recommended for consumption. The golden waxy cap is small and scarce; it is rarely found in abundance.

FUN FACT: This mushroom grows throughout Europe, Germany, Poland, and Switzerland, and is on the endangered list in those places. But in the United States it can be quite common and is considered an everyday woodland mushroom.

HEDGEHOG

COMMON NAME	SCIENTIFIC NAME
Hedgehog	*Hydnum repandum*

ORDER: Cantharellales

FAMILY: Hydnaceae

This is a very fun mushroom to find! It is cute like a hedgehog and even bears a bit of a resemblance to one—hence the name. These mushrooms are a cream or tan color and have tiny spine-like hairs on the bottom that contain the spores. When young, this mushroom has a crease or divot on the top of the cap; as it matures it will flatten out. It has a small, short stem with a big cap that tends to be wavy or irregular. The top of the cap is smooth and almost porous looking, and when bruised it can stain yellow to brown. The older mushrooms are known to have more of a bitter taste, so they are not recommended as much for cooking. However, the younger, softer ones are foraged regularly and sold commonly all over the world.

HABITAT: These mushrooms may be found along the North Atlantic coast, in dense coniferous forests, under leaf litter and can sometimes be found in fairy rings.

DIFFICULTY LEVEL: These mushrooms are rare, but can be collected in abundance when found.

TYPICAL SIZE: The stem is 1 to 4 inches long and 1¼ inch thick; the cap can get up to 10 inches wide, but is typically measured around 6½ inches wide.

NOTES: Hedgehog mushrooms can grow singly or in groups. Once you spot one, it is easier to find others nearby or in groups. They also contain a small amount of oil which suspends their spores before being released.

CONSERVATION STATUS: None.

LOOKALIKES: The most common hedgehog lookalike is hydnum rufescens, which is a bit tanner and more cream colored. It also has a smaller body, with underbelly spines that are separate from the stem.

EDIBILITY AND USES: Commonly sold and used for culinary purposes. Hedgehog mushrooms can be used in place of chanterelles; they are super tasty and carry a similar texture and taste. They can be preserved and pickled for a long time, and are traditionally pickled and stored in climates with harsh winters and intense weather. Hedgehog mushrooms are delicious when sautéed in butter or oil and put on top of bread or crackers. They also go well in pasta dishes and stews.

FUN FACT: This is a very fun mushroom to find! It is cute like a hedgehog and even resembles one a little bit. The underbelly of the hedgehog mushroom has tiny spine-like hairs that contain the spores.

HEN OF THE WOODS

COMMON NAME
Hen of the Woods

SCIENTIFIC NAME
Grifola frondosa

ORDER: Polyporales

FAMILY: Meripilaceae

Hen of the woods, otherwise known as maitake, grows at the base of most trees, similarly to chicken of the woods (Laetiporus sulphureus). It also has a similar polypore structure, fanning outwards and displaying disks of edible mushrooms. These mushrooms are most commonly found on oak trees, and will continue to come up year after year. In North America, maitake can grow up to 5 feet in width. And in Japan, maitake has been reported to grow up to 100 pounds in one fruit. Like many other edible mushrooms, hen of the woods is known for its medicinal properties, carrying various benefits when dried, extracted, or encapsulated.

HABITAT: These mushrooms are found on oak trees in wooded or forested areas along northeastern parts of the United States. Maitake is a late summer mushroom, but it may also be found in the beginning of autumn in more temperate climates.

DIFFICULTY LEVEL: These mushrooms are a fairly rare find.

TYPICAL SIZE: The fruiting body of this mushroom can span anywhere from 3 feet (which is typical) all the way to 5 feet (which is rarer but possible).

NOTES: These mushrooms are known in Asian folk medicine or traditional Chinese medicine (TCM), and can assist with controlling high blood cholesterol, type II diabetes, reproductive system cancers (such as breast or prostate), cold and viral infections, and autoimmune diseases. When dried and extracted, it can be taken in a tincture, powder, or encapsulated to be safely ingested as a daily supplement.

CONSERVATION STATUS: None.

LOOKALIKES: Chicken of the woods is a common lookalike, with the mushroom growing in a similar spot on the base or trunk of trees, and fanning out in a large display of edible mushrooms. However this mushroom has a bright orange color instead of the gray with black tips that maitake displays.

EDIBILITY AND USES: This mushroom is edible and can be foraged for both edible and medicinal uses. It has a long history of culinary use all over the world, and is especially prized as a delicacy in particular parts of Japan where it can be enjoyed in many dishes.

FUN FACT: These mushrooms are used in traditional Chinese medicine (TCM) to assist with high blood cholesterol, type II diabetes, reproductive system cancers (such as breast or prostate), cold and viral infections, and autoimmune diseases. They can be eaten in dishes to get their full polysaccharide benefits, but they can also be concentrated for potent extraction and ease of use.

HONEY

COMMON NAME
Honey

SCIENTIFIC NAME
Armillaria mellea

ORDER: Agaricales

FAMILY: Physalacriaceae

The honey mushroom is a cute little mushroom that is found in clusters on the crevices and knots on the bottom or around the stumps of trees. Although it is often an edible mushroom, some people may have an allergy or intolerance to it. It is also known to produce light through bioluminescence in its mycelium strands. So if foraging at night, these should hold a faint glow, allowing a beautiful display of spores under the cap catching the wind, if you are lucky. These mushrooms are also known to be plant pathogens and may cause root rot in the trees they grow on. They are often found in the northern hemisphere, primarily on hardwood trees and other decaying or dead wood in the area.

HABITAT: Normally found on hardwood trees throughout North America, growing on the bases of the stumps or around the trunks.

DIFFICULTY LEVEL: These mushrooms can be a bit harder to find; although they are not rare or endangered, they can be a difficult forage.

TYPICAL SIZE: The cap of this mushroom is 1 to 6 inches across, and the slender stem is about 1 inch in width and up to 8 inches long.

NOTES: The honey mushroom can be found with a wet film covering its entire body, which can be tacky or sticky when still wet. The mushroom (like its name) resembles the color of rich honey; it can have the appearance of honey being dropped or spread on top of the clusters. Its gills are a soft white that can tint to a dark yellow when maturing.

CONSERVATION STATUS: None.

LOOKALIKES: The funeral bell mushroom or galerina is a common lookalike, although it is significantly darker in appearance and does not use bioluminescence through its mycelium.

EDIBILITY AND USES: The honey mushroom is a great culinary mushroom with a nutty and sweet flavor—perfect for baking and making pastries. Try just a small portion at first to make sure that you do not have a reaction. Some people have an allergy or intolerance to honey mushrooms.

FUN FACT: You may find the honey mushroom with a wet stickiness covering its entire body. The mushroom resembles the color of rich honey and can have the appearance of honey being dropped or spread on top of the clusters. And the similarities to honey do not end there; the honey mushroom is edible and carries a nutty, sweet taste, making it ideal for use in baking and pastries.

JELLY BABY

COMMON NAME
Jelly Baby

SCIENTIFIC NAME
Leotia lubrica

ORDER: Leotiales

FAMILY: Leotiaceae

Jelly baby is a small mushroom with what looks to be a deflated cap but is actually part of the mushroom's fully fruited stage. Although its name may reflect one of our favorite spreads to put on toast, this mushroom is not commonly used for cooking, mainly because its small size and bland taste would not amount to much. The jelly baby mushroom grows under the soil and pops up in dense clusters on forest floors under conifer trees or around leaf litter and scattered forest mulch. The structure of this mushroom is similar to that of jelly or gel; it can be smooth and squishy all throughout the body and can hold a cooler temperature.

HABITAT: These mushrooms are mostly found in woodland and forested areas in cooler climates throughout North America.

DIFFICULTY LEVEL: Because of its small size this mushroom can be harder to spot, but when found can be found in abundance.

TYPICAL SIZE: Jelly baby mushrooms reach no larger than 2 inches tall with a 1 inch cap.

NOTES: When cut or picked from the ground this mushroom excretes a nontoxic gel that it produces from the inside. Jelly babies are commonly seen by or on moss growing in large clusters; when foraging, I look for clumps of moss or mossy rocks to find jelly babies hiding around nearby or growing on top of the moss.

CONSERVATION STATUS: None.

LOOKALIKES: One of the most common mushrooms mistaken for the jelly baby is Leotia atroverins. It can display almost identically to the jelly baby but has a green cap and slightly thicker stem. This mushroom's edibility is not known so is not recommended for consumption until further research is done.

EDIBILITY AND USES: This mushroom is not commonly used for culinary purposes.

FUN FACT: When cut or picked from the ground, a jelly baby mushroom excretes a nontoxic gel that it produces from the inside. Jelly babies are commonly found growing by or on moss in large clusters.

KING BOLETE

COMMON NAME
King Bolete

SCIENTIFIC NAME
Boletus edulis

ORDER: Boletales

FAMILY: Boletaceae

The king bolete is an edible bolete species that is commonly foraged for culinary purposes. This species tends to grow in densely populated forests that mostly have coniferous trees, creating a synergetic chemistry that can be symbiotic with other living plants and trees in the area. When the fruiting body is fully mature, it has a brown surface extending down the cap that fades to a lighter shade of brown around the rim. It has a polyporus underbelly branching in a tube-like formation instead of gills.

HABITAT: This mushroom can be widely found in various parts of North America in densely populated forests that contain coniferous trees like pine, spruce, and fir trees. It is usually found in the ground on rich soil around pine leaves or scattered leaf litter.

DIFFICULTY LEVEL: This mushroom is a common find for most mushroom hunters.

TYPICAL SIZE: The cap ranges from 10 to 12 inches and can sometimes get up to 16 inches (although rare).

NOTES: This mushroom is known for its culinary fame; many chefs and culinary artists use it in a wide variety of dishes, as it has a rich and delightful taste when cooked. It has long been known all over the world as a seasonal delicacy.

CONSERVATION STATUS: None.

LOOKALIKES: Bolete regineus is a common bolete that many mistake for the king bolete, mainly because they have the same colored cap and stem. But, if flipped over, you can clearly see a bright yellow porous underbelly that is uniquely different from the king bolete. Bolete regineus is also edible and not poisonous.

EDIBILITY AND USES: This mushroom is edible and is known to be used in various dishes around the world including soups, pastas, risottos, stews, and more.

FUN FACT: The king bolete is an edible bolete species that is commonly foraged for culinary purposes. It has a polyporus underbelly branching in a tube-like formation instead of gills, which can help you further identify this mushroom as a true bolete species.

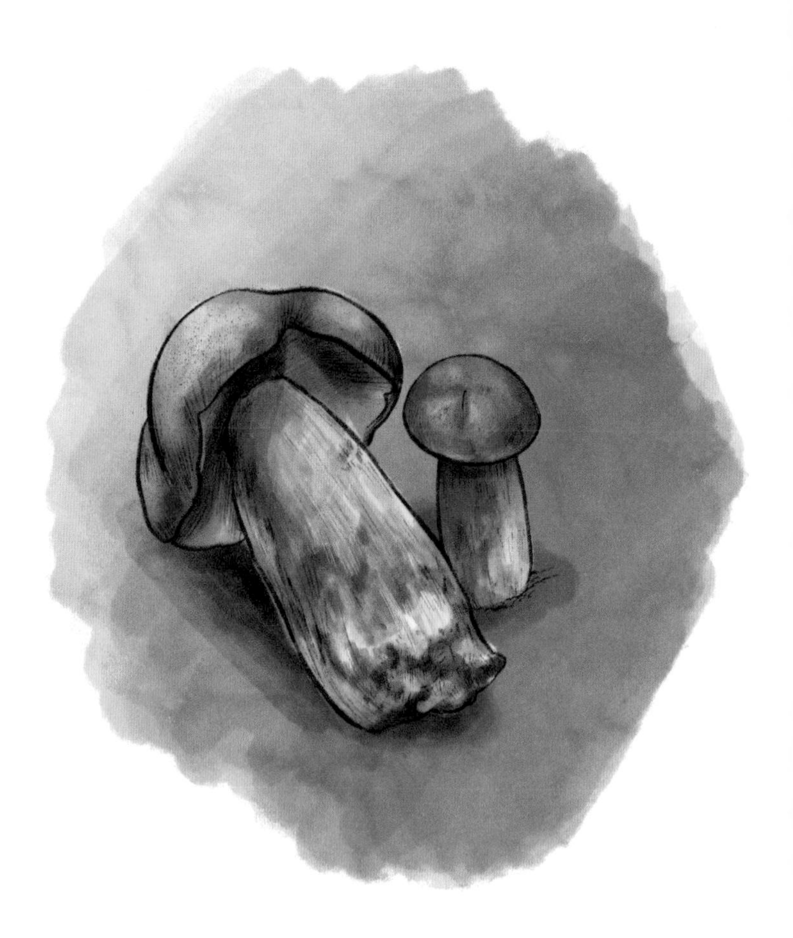

LILAC BROWN BOLETE

COMMON NAME	SCIENTIFIC NAME
Lilac Brown Bolete	*Sutorius eximius*

ORDER: Boletales

FAMILY: Boletaceae

The lilac brown bolete is a common bolete that is rich dark brown in appearance, with a rust colored underbelly that is porous like the king bolete. It is commonly found here in North America but has also been found throughout the world. It is an edible variety and can be found in many field guides and foraging books. It can vary in color, presenting with a slighter brown, but is most commonly known to be a rich coffee brown.

HABITAT: This mushroom can be found in leaf litter throughout the forest floor. They grow singly but can fruit many in the same area, and can be most commonly seen growing under conifer trees.

DIFFICULTY LEVEL: These are another more common and easy find once stumbling upon a patch. Seeing one can lead to finding a plethora scattered throughout the area.

TYPICAL SIZE: The lilac brown bolete ranges in size from

2 to 4 inches across the cap, with a stem that is 1 to 3 ½ inches long.

NOTES: This mushroom can have a tacky surface when touched, and it bruises dark brown when cut or damaged. When sliced in half, this mushroom displays a solid white and edible flesh that is great for use in cooking. Similar to the king bolete, this mushroom has a symbiotic relationship with the plants and trees living nearby or in the area, creating a synergistic relationship with the soil and plants.

CONSERVATION STATUS: None.

LOOKALIKES: Tylopilus Violatinctus is often mistaken for a lilac brown bolete, because it resembles a bolete and is sometimes known as the violet bitter bolete. But the color is a lot lighter than lilac brown, and its stem is thinner and longer. The lilac brown bolete is much thicker and bulbous around the base with a flatter and wider cap.

EDIBILITY AND USES: This mushroom is known to be edible and used around the world in many different ways. It can be very rare, but in some cases this mushroom has been reported to produce intoxicating or poisonous reactions in people when eaten, so precaution is always heavily advised.

FUN FACT: You may find that this mushroom has a tacky or even sticky surface when touched, and it bruises dark brown when cut or damaged. When sliced in half, the lilac brown bolete displays a solid white and edible flesh that is incredibly tasty.

LION'S MANE

COMMON NAME
Lion's Mane

SCIENTIFIC NAME
Hericium erinaceus

ORDER: Russulales

FAMILY: Hericiaceae

Another all-time favorite on my list of mushrooms is the majestic lion's mane! This beautiful and medicinal mushroom is known to be a comb tooth fungus. It carries specific traits like long, cone-like, thick, and shaggy white hairs that cover the mushroom. These hair-like extensions are actually the flesh and exterior fruiting body of this mushroom, and are quite edible. Lion's mane can be confused with other mushrooms in the same family, but its size and shape is quite different. The color can range from light pink (when very young) to a more cream-colored white (when fully matured and starting to die off).

HABITAT: Lion's mane mushrooms are typically found growing on decaying or dead hardwood trees (such as beech wood), but they can also form on living trees. They can be found in the late summer months across North America, especially on the eastern side.

DIFFICULTY LEVEL: Depending on region and climate, lion's mane mushrooms can be common and easy to find. Look for a bright white shaggy mushroom dangling from a dead tree knot or branch.

TYPICAL SIZE: These mushrooms can get pretty big when growing in the wild or being cultivated indoors. The fruit bodies can typically get anywhere from 2 to 16 inches at full maturity. The long dangling icicle-like cones hanging from the fruit body can get up to ½ inch long.

NOTES: This mushroom is heavily used in cooking as a meat replacement for seafood and other meats similar to its texture. It carries a variety of medicinal benefits that some cultures have used for many years to help aid in memory and cognitive function. This mushroom can be finely powdered and processed producing a fluffy powder that be added to soups, smoothies, encapsulated, or made into a tincture.

CONSERVATION STATUS: In some countries this species is red listed as a threatened species, but in the United States there is no known conservation status.

LOOKALIKES: Hydnellum caeruleum are also white like lion's mane mushrooms; however, unlike lion's mane, the toothlike structure is on their underbellies (not on their fruiting body) to expel their spores. They also grow on the forest floor or on fallen logs, whereas lion's mane likes to grow higher up on trees and branches.

EDIBILITY AND USES: This mushroom is one of the top choices among chefs and food enthusiasts all over the world! Lion's mane got a huge trend hit in 2019 and 2020 when Paul Stamets appeared on a TED talk explaining the power of mushrooms. In that same timeframe, he also came out with the movie *Fantastic Fungi*, and launched a line of lion's mane powders and tinctures—starting a huge buzz and craze in the world of mushrooms. Lion's mane is now one of the most heavily manufactured and used mushrooms in the market of health and wellness.

FUN FACT: Lion's mane carries a variety of medicinal benefits that some cultures have used for centuries, most commonly to aid memory and cognitive function for people suffering from dementia, Alzheimer's, and other neurological disorders. This mushroom can be finely powdered and processed; it produces a fluffy powder that is delightful when added to soups or smoothies. The powder can also be encapsulated or made into a tincture.

LOBSTER

COMMON NAME	SCIENTIFIC NAME
Lobster	*Hypomyces lactifluorum*

ORDER: Hypocreales

FAMILY: Hypocreaceae

Lobster mushrooms are an interesting species of parasitic fungi that attack the exterior of certain mushrooms, turning them a bright red-orange color like a lobster, and can sometimes have the appearance of a lobster tail. They are commonly used for culinary purposes and can be found in markets worldwide; very few have been sold in grocery stores because it is more of a seasonal forage.

HABITAT: These mushrooms are typically found on the host of milkcaps and brittlecaps in lush and fertile forests that have lots of coverage and overhanging trees. You might find them in forests in the United States and Canada.

DIFFICULTY LEVEL: These mushrooms are more common to find in Oregon and along the Pacific Northwestern coast and in old growth forests.

TYPICAL SIZE: These mushrooms can grow to be 6 to 8 inches tall.

NOTES: As stated above, this mushroom is in fact not a mushroom at all! It is a parasitic fungi that lives by attacking the exterior of other edible mushrooms and making it unrecognizable, resembling that of a lobster tail. Because of this, most foragers will not be able to identify what mushroom is being taken over by this fungi, and the host mushroom could be poisonous or deadly. So, always take proper precautions, go with a buddy when foraging, and always identify the mushroom that is infected before consumption (unless bought from a trustworthy market).

CONSERVATION STATUS: None.

LOOKALIKES: None.

EDIBILITY AND USES: This mushroom is commonly used in cooking, and is another choice edible for gourmet dishes and meat substitutes because of its seafood-like flavor. The safest way to eat this mushroom is to buy it from experienced foragers or mushroom growers, and to cook it thoroughly. This parasitic fungi attacks mostly edible species of mushrooms, but a few have been reported poisonous.

FUN FACT: The lobster mushroom is technically not a mushroom, but rather a parasitic fungi that lives by attacking the exterior of other edible mushrooms and making them unrecognizable. The result of this parasitic takeover often ends up resembling a lobster tail. Because of this, most foragers will not be able to identify what mushroom is being absorbed by the lobster fungi, so beware: the host mushroom could be poisonous or deadly. Because of this, foraging lobster mushrooms is not recommended. If you want to try them, they are commonly sold through markets and by mushroom cultivators and expert foragers.

MATSUTAKE

COMMON NAME
Matsutake

SCIENTIFIC NAME
Tricholoma matsutake

ORDER: Agaricales

FAMILY: Tricholomataceae

The undeniably cute matsutake is a great mushroom prized in Asian cooking. It is known to grow abundantly in China and Eastern Asia as well as in parts of Europe, Sweden, and North America. Its main shape is round and long, resulting in small stout caps with thick rounding stems. Matsutake are usually hiding under scattered leaf litter, pines, or mulch on the forest floor. They are stout with broad and thick stems, sometimes with dark brown markings along the stipe (stem) with a dark brown cap.

HABITAT: These mushrooms are found in various countries and the United States, where they are commonly found growing on forest floors under a layer of leaves that have fallen from old growth trees.

DIFFICULTY LEVEL: These mushrooms are a rare find when foraging.

TYPICAL SIZE: The cap of the mushroom is 2 to 8 inches across and the stem is from 2 to 6 inches tall.

NOTES: Matsutake has a special relationship to partnering plants and trees nearby, creating a symbiotic relationship with the soil and roots. This creates an atmosphere of symbiotic relationships working together to keep the area healthy and rich in biodiversity and fertility.

CONSERVATION STATUS: None.

LOOKALIKES: Matsutake can be mistaken for shiitake when young. This is a common misconception; however, they can easily be differentiated because they grow in vastly different habitats.

EDIBILITY AND USES: Matsutake is known for its culinary purposes and is used in a lot of recipes. In fact, this mushroom is delightfully named after a dish that made it famous, matsutake Gohan, which translates as wild pine mushroom rice.

FUN FACT: Matsutake is a precious mushroom. When harvested in Japan, they can cost more than five hundred dollars per kilogram. It is difficult for matsutake foragers to predict how well they will grow. And, since the 1950s, the Japanese matsutake harvest has dropped more than 95 percent, driving the price sky high.

MOREL

COMMON NAME
Morel

SCIENTIFIC NAME
Morchella esculenta

ORDER: Pezizales

FAMILY: Morchellaceae

Morels are a great and rare to find mushroom, making these a top hit on most mushroom hunter's bucket lists. Sadly, I have still yet to come across an abundant morel patch. But these are a true treasure to be found; like mining for gold these can definitely be considered a mushroom hunter's gold mine. Morels have a special relationship with trees in the area. Fruit trees like apple or ash trees are known to be the most commonly found around morels. In cooler places they tend to pop up from May to June, but in more temperate climates they can be seen in spring from February to June. They are particularly sought after for their rich culinary history. Whether used dried or cooked fresh, these mushrooms have a magnificent taste packed with flavor. For such a tiny mushroom it has a big presence in the culinary world. Mushroom hunters will collect pounds of these for months, foraging in specific places they have gone to year after year to make a living off just one season. They are expensive because they can only be foraged seasonally.

HABITAT: Morels are commonly found around apple and ash trees. They can be found from California up to Washington; throughout Nevada, Arizona, and Texas; and all throughout the east coast of North America.

DIFFICULTY LEVEL: Commonly found in abundance.

TYPICAL SIZE: A small, stout mushroom with an elongated cap measuring around 3 to 5 centimeters tall in full length.

NOTES: Morels are sometimes disfigured honeycomb shapes. When growing to full maturity they can be easily mistaken for false morels or a different mushroom altogether. Morels do not grow in clusters, but rather in abundant patches under fertile trees.

CONSERVATION STATUS: None.

LOOKALIKES: False morel mushroom Gyromitra esculenta is commonly confused with true morel mushrooms. The appearance is similar to the morel, yet the cap is more disfigured with a different pattern of creases and folds in the cap. The coloring can also be slightly different; the false morel mushroom has an orange or reddish cap compared to the yellow (in early stages) and gray to black color of true morels.

EDIBILITY AND USES: Morels are widely used for their culinary excellence and rich history in recipes. They are great in soups, stocks, gourmet dishes, and more.

FUN FACT: Morels are a true treasure waiting to be found. They can definitely be considered a mushroom hunter's gold mine! They are widely used for their culinary excellence and rich history in recipes for soups, stocks, and more. In the Southern United States, they are often dipped in flour, breading, or batter, and fried. Morels only appear seasonally, so the price for foraged morels is often quite high.

MOUNTAIN MOSS PSILOCYBE

COMMON NAME	SCIENTIFIC NAME
Mountain Moss Psilocybe	*Deconica montana*

ORDER: Agaricales

FAMILY: Strophariaceae

Like its name suggests, mountain moss psilocybe is a mushroom that grows on moss in high altitude mountain ranges. Commonly found throughout North America, this mushroom has a small, slender body with an umbrella-like head. Hued a reddish brick brown color, it pops on moss-ridden rocks and mountains. This mushroom has been seen growing singly in less densely populated areas and in small groups on heavily moss covered rocks.

HABITAT: You will find mountain moss psilocybe on high rocky mountain ranges with moist soil and dense moss areas. You can also catch them in dunes and among pine trees. This mushroom has been spotted worldwide. In North America, it is frequently found in California, Oregon, Washington, and other places along the Pacific coast.

DIFFICULTY LEVEL: Common to see.

TYPICAL SIZE: The cap is 0.5 to 1.5 centimeters across, and the stem measures 1.5 to 4 centimeters long.

NOTES: Although the name makes it sound like a hallucinogenic species, this mushroom is rarely reported to be one. Regardless of its name, it is simply a small, thin mushroom that likes to grow in mossy patches in rich, lush forests.

CONSERVATION STATUS: None.

LOOKALIKES: Psilocybe strictipes is a lookalike, although this mushroom can grow slightly taller, is browner in color, and has more of a peak at the top on the cap. This mushroom is indeed hallucinogenic and can impair the consumer for a few hours, but it is not poisonous.

EDIBILITY AND USES: This mushroom is inedible but not enough is known about its toxicity level. It is not recommended for consumption.

FUN FACT: Although the name mountain moss psilocybe makes it sound like a hallucinogenic species, this mushroom is rarely reported to actually be one. It is simply a small, thin mushroom that likes to grow in mossy patches in rich, lush forests.

OREGON WHITE TRUFFLE

COMMON NAME
Oregon White Truffle

SCIENTIFIC NAME
Tuber oregonense

ORDER: Pezizales

FAMILY: Tuberaceae

The white Oregon truffle is an edible truffle known for its rich taste in modern culinary arts and cooking. They are a creamy white with outlines of brown around the crevices and creases of the truffle. They have a unique earthy and cheesy taste, and can be infused in many oils and butters for a white truffle flavored spread or cooking oil. They have to be dug up from the ground, and are known to be growing under Douglas fir trees and other old conifers.

HABITAT: These truffles are typically found in dense old growth forests under the surface of Douglas firs or near the root structures. They can be most commonly found in Oregon and in patches. They have also been found in Washington.

DIFFICULTY LEVEL: More of a difficult find due to having to dig and find them underground without true markers.

TYPICAL SIZE: White Oregon truffles can range from quite small (1 inch) to moderately large in size (around 5 inches, or the size of a small lemon).

NOTES: This mushroom is known for its distinct taste and aroma, which can span from a light cheesy flavor to a rich earthy and deep taste with age and preparation.

CONSERVATION STATUS: None.

LOOKALIKES: Elaphomyces granulatus, also known as false truffle fungus, can be mistaken for Oregon white truffles. False truffles typically have small mushrooms growing alongside them, whereas Oregon white truffles do not; instead, they have a symbiotic relationship with the rootzones of certain trees. False truffles are completely black inside and are highly poisonous; more research needs to be done on toxicity levels.

EDIBILITY AND USES: Widely used in gourmet cooking, this mushroom is highly edible and a great addition to dishes. But it is not to be mistaken with its false lookalike, which can be deadly or very poisonous! Always buy from trusted growers or foragers when ingesting mushrooms or have a proper way to identify the mushroom.

FUN FACT: Some of the most dedicated truffle hunters have a secret weapon. And while I may get flack for it I will expose it here: They use truffle dogs! Many breeds of dogs can be trained to use their amazing sense of smell to sniff out truffles' hiding spots. Watching a truffle dog work is an amazing thing to see—and it is a fun activity for working dogs!

OYSTER

COMMON NAME
Oyster

SCIENTIFIC NAME
Pleurotus ostreatus

ORDER: Agaricales

FAMILY: Pleurotaceae

Oyster mushrooms are commonly found throughout North America, Japan, and Germany. It is a more sought out mushroom for it has a safely edible body that is rich in nutrients. It commonly grows on trees in the shape of an oyster or a clam and can grow significantly big making it easy for spotting and a bountiful forage. Oyster mushrooms are also grown commercially all over the world for their easy and versatile cultivation methods. They can be found in markets, grocery stores, and are easily foraged when properly identified.

HABITAT: These mushrooms live in tropical regions all over the world. They prefer coastal areas that have old beech trees or decaying soft wood. These mushrooms have been spotted across the United States, especially along the Pacific and North Atlantic coasts.

DIFFICULTY LEVEL: Very common.

TYPICAL SIZE: These mushrooms have broad round cups that fan outward up to 11 inches wide. The stem protruding out from the tree varies in size and width due to how the mushroom is positioned on the tree.

NOTES: They grow outward in a vertical flat-like shape and can typically grow in clusters scattered throughout the trunk of the tree. They are also known to be able to absorb toxic chemicals, as well as oil and petroleum. Using special engineering, mycologists, scientists, ocean biologists, and wildlife conservationists can work together to create new technologies that clean up oil spills and other damaging toxins that may lead to declines in wildlife and natural habitats.

CONSERVATION STATUS: None.

LOOKALIKES: The jack-o'-lantern mushroom can look quite similar to a pink oyster mushroom, but the color and shape may vary. There are observable differences between them, like location: oysters like growing on trees, whereas jack-o'-lantern mushrooms grow from forest floors and mulch. Also notice how they grow: oysters like to grow in a cluster and branch out further from the stem, whereas jack-o'-lantern mushrooms grow in one big group and have individual mushrooms with a full cap and stem growing singly in the group.

EDIBILITY AND USES: This mushroom is commonly used in culinary and gourmet cooking. It can be used as a meat replacement, and is often seen in the produce section at most grocery stores. It is commonly sold through mushroom growers, and has been used medicinally in a few cultures.

FUN FACT: Oyster mushrooms have an amazing ability to absorb toxic chemicals, such as oil and petroleum. They can also be used to safely and efficiently clean toxic particles out of water runoff.

Mycologists, special engineering scientists, ocean biologists, and wildlife conservationists can work together to create new technologies that clean up oil spills and other damaging toxins that may lead to declines in wildlife and natural habitats.

PARROT

COMMON NAME
Parrot

SCIENTIFIC NAME
Gliophorus psittacinus

ORDER: Agaricales

FAMILY: Hygrophoraceae

This mushroom is a cute and colorful one! When growing, it can have a waxy or wet appearance and is a light green; as it ages, the green will turn to a yellowish orange or even a light shade of pink. Its gills are also green fading out into a yellow around the rims. It grows in small groups of two or three—sometimes four at most.

HABITAT: It is commonly found in wet, grassy fields and near streams or light bodies of water in the Pacific Northwest and throughout the northern coast of the United States, but some have been found in southern Ontario in Canada. Interestingly, this mushroom has been spotted in European grasslands as well.

DIFFICULTY LEVEL: Rare to find.

TYPICAL SIZE: This mushroom is quite small. The cap gets no bigger than an inch across and the stem can get to about 3 inches high.

NOTES: The parrot mushroom tends to prefer dead grass and roots to attach onto when growing. Mosses are also seen growing around or with this mushroom, showing a symbiotic relationship between the two.

CONSERVATION STATUS: None.

LOOKALIKES: Hygrocybe chlorophana, or golden waxy cap, can be commonly mistaken for the parrot mushroom when young and tiny, but has a distinct yellow color throughout its whole body that does not fade to green or create different colors when fully matured.

EDIBILITY AND USES: This mushroom is considered edible, but when eaten in large amounts (over twenty mushrooms) can lead to gastrointestinal issues.

WARNING: ***Be careful. Twenty of these mushrooms may not seem like a large quantity when cooking due to their small size.***

FUN FACT: Parrot mushrooms are a spectacular sight. Their colorful display mainly consists of shades of green, but can fade into a more multicolored appearance. Some dare to eat them, but because of their small size and tendency to cause gastrointestinal issues, it is best to simply view them from afar and admire their colors.

POTENT PSILOCYBE

COMMON NAME	SCIENTIFIC NAME
Potent Psilocybe	*Psilocybe azurescens*

ORDER: Agaricales

FAMILY: Hymenogastraceae

Psilocybe azurescens is a psychedelic mushroom containing the compounds psilocin and psilocybin. It is known to be found in abundance throughout northern California and up to Oregon. This mushroom has a tannish or orange cap protruding into a nipple at the top, with a cream or brick colored stem than can become bruised with blue stains. When picked and handled, this mushroom will easily stain blue due to the entheogenic compounds and oxidation that is occurring.

HABITAT: These mushrooms can be found growing in wood chip piles and mulch-littered forests with heavy bark sheddings throughout the Pacific Northwest, from Washington down to California.

DIFFICULTY LEVEL: Commonly found, but illegal (felonious) to harvest in large amounts in some countries, including the United States.

TYPICAL SIZE: The cap measures a little over 3 inches in width and the stem can get up to 7 inches long.

NOTES: These mushrooms are one of the most potent strains; take caution if using for therapeutic purposes. The mushrooms grow singly in patches and can also be found growing in hay or near dead grass. They have a symbiotic relationship to wood, growing mainly on wood chips that have been untreated in parks and in forest mulch.

CONSERVATION STATUS: None, but is illegal in some countries.

LOOKALIKES: Psilocybe semilanceata, otherwise known as the liberty cap, is another psychedelic mushroom that this mushroom can be commonly mistaken for. The specific and very noticeable differences are in the visual appearance of the mushroom. Psilocybe azurescens have a more elongated cap, whereas psilocybe semilanceata has a shorter, flying saucer-shaped cap. Liberty caps also tend to have a wavy stem making an *s*-like formation while growing.

EDIBILITY AND USES: Inedible, but may be used for psychedelic therapy or therapeutically through other practices.

FUN FACT: These mushrooms are one of the world's most potent strains of entheogenic fungi and should be handled with caution (and only for therapeutic purposes in legal states). Entheogenic fungi are also known as "magic mushrooms" or therapeutic mushrooms. They contain a compound called psilocybin that gives magic mushrooms their famous psychedelic effect. Other compounds like psilocin are also active in potent psilocybe and other mushrooms that are entheogenic; they also have a similar psychoactive effect to LSD and DMT. Magic mushrooms have a place in the folklore of many cultures throughout the world. They've also been used in ancient cacao ceremonies, found buried with royalty or shamans, and have been a sacred tool and medicine to many people.

RED-BELTED POLYPORE

COMMON NAME	SCIENTIFIC NAME
Red-Belted Polypore	*Fomitopsis pinicola*

ORDER: Polyporales

FAMILY: Fomitopsidaceae

This mushroom, like its name, has an identifiable stunning red belt expanding around its polyporus body. This shelf-like mushroom is a hidden gem because it is very medicinal and can be used for a variety of modern ailments. It is said that the Native Americans first used it for tinder to help keep their pipes lit. They also used it to staunch blood, and to prevent or treat headaches.

HABITAT: Typically found in soft and hard wood forests on decaying trees, it is native to North America and is commonly found from California to Alaska, and across British Columbia.

DIFFICULTY LEVEL: Very common.

TYPICAL SIZE: This mushroom can get up to 11 inches wide and can create a 5 inch thick body of rings.

NOTES: The red-belted polypore is in the bracket family and grows in the wounds or knots of trees, creating a fan-like shelf with thin rings and one band of red around its body. It has a brown top fading out into a deep red around the edges, with a porous white underbelly. The red-belted polypore is a slow grower that can take many years to fully reach maturity, adding a new ring each year.

CONSERVATION STATUS: None.

LOOKALIKES: Reishi is a similar lookalike but has a shiny crimson red top that is quite distinguishable from the red belt around the body of the red-belted polypore.

EDIBILITY AND USES: This mushroom is not considered edible but is used in various cultures for medicinal purposes.

FUN FACT: Red-belted polypores have been used in homeopathic medicine for centuries. They are known for their anti-inflammatory benefits. They contain a natural steroid that may help reduce autoimmune disorders and the inflammation they cause.

REISHI

COMMON NAME
Reishi

SCIENTIFIC NAME
Ganoderma lucidum

ORDER: Polyporales

FAMILY: Ganodermataceae

This is another mushroom famous for its medicinal properties, known as "the mushroom of longevity" in traditional Chinese medicine. Another name for it is lingzhi. It has long been a prized mushroom to put in soups, stocks, tinctures, and for medicinal use. Heavily cultivated and foraged for its beauty, this mushroom is a striking shelf display of autumnal colors from deep red to orange, yellow, and white. It can be found on maple and deciduous trees near the base.

HABITAT: Although Asia is the home of the reishi mushroom, it can also be foraged in North America on hardwood and maples trees near the bottom or the stump. It has been found in Ontario, Quebec, New Brunswick, Prince Edward Island, Northern California, Vermont, New York, and Pennsylvania.

DIFFICULTY LEVEL: Extremely rare to find.

TYPICAL SIZE: These mushrooms can get up to 3 to 5 inches across in diameter.

NOTES: This mushroom tends to only take to two or three out of ten thousand trees, so it can be extremely rare and hard to find. It is now being successfully cultivated all over the world in abundance, and can be grown indoors with ease. A variation has occurred due to cultivation techniques, producing more antler-like formations at the top due to high CO_2 and low oxygen levels.

CONSERVATION STATUS: None.

LOOKALIKES: Artist's conk and red-belted polypore are both easy to mistake for reishi. Artist's conk has a tougher and flatter top with a soft polypore underbelly that stains black when scratched. Red-belted polypores are also similar in appearance but grow significantly smaller in size and display a beautiful red belt around the exterior of their body.

EDIBILITY AND USES: Reishi mushrooms have been heavily used in holistic and cultural remedies all over the world for their powerful medicinal benefits.

FUN FACT: Reishi is another celebrity in the mushroom world. This mushroom gets the name "the mushroom of longevity" for a reason! Known in traditional Chinese medicine to bring long lasting health, boost stamina, and build a stronger immune system, it has been around for centuries. Reishi teas and herbal concoctions may aid in physical and mental health. Reishi is a safe adaptogen that can be taken by pretty much anyone to help fight off inflammation, relieve stress (both physical and mental), nourish the body, and help tone the nervous system.

SCALY TOOTH

COMMON NAME
Scaly Tooth

SCIENTIFIC NAME
Sarcodon imbricatus

ORDER: Thelephorales

FAMILY: Bankeraceae

This mushroom is a stout brown mushroom that resembles a burnt marshmallow or scorched wood. It pops up around autumn and is found throughout North America. It grows in single patches and is known to be edible. Some say they have a similar species that is bitterer but this one is quite sweet in taste. This mushroom is related to that in the hedgehog mushroom family and has small teeth on its underbelly instead of gills.

HABITAT: Commonly spotted throughout North America, this mushroom is found in rich and damp soils that have been populated by fir trees.

DIFFICULTY LEVEL: Easy and common to find.

TYPICAL SIZE: The stem measures 4 inches high, the cap has a width of 12 inches, and the underbelly teeth structures measure about 1.5 centimeters long.

NOTES: These mushrooms can be used for culinary purposes, and are some of the more commonly found species of foraged and edible mushrooms in Northern California. While some can be particularly bitter, they are all considered safe within this specific species.

CONSERVATION STATUS: None.

LOOKALIKES: Strobilomyces strobilaceus is a mushroom that looks similar to the scaly tooth but has a taller and more elongated stem with a shaggy texture. It has a number of conjoined toothlike gills underneath it with a similar pattern on its cap. Strobilomyces scabrosus, also known as old man of the woods, is another similar type of mushroom with little edibility that has a similar pattern to the scaley tooth. If looking closely you can see that the pattern is actually small bits of brown tufts on top of a cream colored cap.

EDIBILITY AND USES: Most commonly found to be edible although taste may vary from bitter to a more pleasant and sweet taste.

FUN FACT: The scaly tooth mushroom resembles a burnt marshmallow or scorched wood. It grows in single patches and is known to be edible. It is related to the hedgehog mushroom family and, like the hedgehog mushroom, it has small teeth on its underbelly instead of gills.

SHAGGY MANE

COMMON NAME	SCIENTIFIC NAME
Shaggy Mane	*Coprinus comatus*

ORDER: Agaricales

FAMILY: Agaricaceae

The shaggy mane mushroom is edible when young and rather fresh. It has the ability to completely dissolve into an inky blob within a few hours of being picked. A bright white mushroom with an elongated cap with little shags of its exterior around it, it grows scattered throughout leaf litter. When picked and held it can disperse its spores through an inky fluid that gathers under the cap and quickly disintegrates. Although if properly prepared, it can be stored for cooking and can be enjoyed a few hours after it is foraged.

HABITAT: These mushrooms have been found in northwest territories of Canada, and from Alaska down to Northern California. However, many have been spotted in the eastern part of the United States, and even down to Mexico. They hide in leaf litter and in grasses and fields with rain or foggy microclimates.

DIFFICULTY LEVEL: Very easy and common to find.

TYPICAL SIZE: The stem can get up to 3 inches high, with a cap measuring 1 to 2 inches across.

NOTES: This mushroom displays a very interesting feature when releasing spores or just being picked, as it turns itself pitch black and fully dissolves.

CONSERVATION STATUS: None.

LOOKALIKES: Magpie fungus Coprinopsis picacea can be easily mistaken for the shaggy mane because of its similarity in size, shape, and patterning. Its fuzzy shaggy body is similar in appearance, but this mushroom is browner and more egg-like with white tufts. Shaggy mane mushrooms have a more elongated bright white cap and stem.

EDIBILITY AND USES: This mushroom is known to be edible but must be eaten a few hours upon finding it, or be prepared appropriately for storage. It does not hurt to eat desintegrated shaggy mane mushrooms, but they can be slimy and less flavorful than those with fresh, solid flesh.

Keep shaggy manes dry and be gentle with them, as water and bruising or cutting make them dissolve faster. In Vietnam, people sometimes transport them in the hollows of empty egg cartons.

An avid mushroom lover might even carry a small skillet and butter in their backpack so they can enjoy shaggy mane mushrooms the instant they are discovered.

FUN FACT: When young, shaggy mane mushrooms have elongated white caps and matching white stems. As they grow older and are stepped on or picked, they can disintegrate within minutes, leaving behind nothing but a black inky blob that contains their spores. In the 1700s, liquified shaggy mane mushrooms were used as writing ink.

TINDER POLYPORE

COMMON NAME	SCIENTIFIC NAME
Tinder Polypore	*Fomes fomentarius*

ORDER: Polyporales

FAMILY: Polyporaceae

The tinder polypore is a parasitic fungi that infects trees, and grows a conk-like protrusion out of the broken bark or wound of the tree. The tinder polypore, like its name suggests, is known for being used as kindling for fire. This mushroom is very light and, because of its polypore structure, dries out fairly quickly after being detached from the tree. This makes it great to use for tinder, and has been used as such by nomadic or traveling cultures and tribes. Tinder polypore's shape can resemble a cow or horse hoof, and it varies in color from a light shade of gray to a dark brown.

HABITAT: These mushrooms are found throughout the northeast of North America, namely from Southern Ontario to Prince Edward Island. In the United States, they can be spotted in Maine, Vermont, Massachusetts, and other east North American climates. They grow on the same hardwood tree in the same spot year after year. They can be found in drier and warmer climates, too (like Africa), around 80 to 86 degrees Fahrenheit.

DIFFICULTY LEVEL: More common to find.

TYPICAL SIZE: This mushroom is around 2 to 7 inches across and can get up to almost 10 inches wide and thick.

NOTES: This mushroom is known for its tinder and medicinal use, but it is not considered an edible mushroom. Many known mycologists and travelers would carry pieces of this mushroom when hunting or living in the woods foruse in the creation and maintenance of fires.

CONSERVATION STATUS: None.

LOOKALIKES: Fomitopsis pinicola is another dead wood conk or polypore that resembles the tinder polypore in formation and size, but the color is a rich red or deep orange hue.

EDIBILITY AND USES: The tinder polypore is not edible but is used medicinally in many cultures. This mushroom is covered in a rough skin, which, when boiled and strained, can be used in soups, stocks, and medicinal syrups and tinctures. It can also be dried out for use in powder form.

FUN FACT: For centuries, tinder polypores have been collected for use as fire starters. Making a fire from nothing can be extremely difficult, so the fact that this mushroom can start and keep a fire going is incredibly resourceful. In ancient times, some travelers would hang pieces of it from necklaces worn during long excursions.

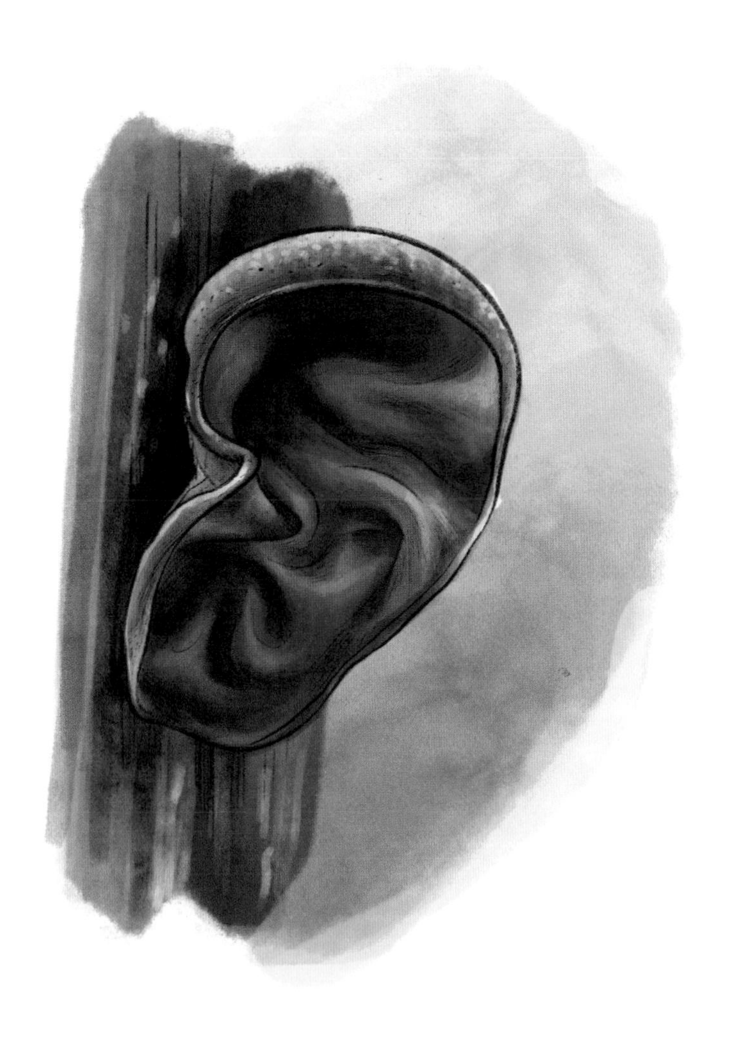

TREE EAR

COMMON NAME
Tree Ear

SCIENTIFIC NAME
Auricularia auricula-judae

ORDER: Auriculariales

FAMILY: Auriculariaceae

The forest is listening . . . this fun mushroom is shaped like an ear! It is also known as black wood ear and jelly ear fungus. It is an edible fungus that is popularly used in Chinese medicine and classic dishes like hot and sour soups. Tree ear mushroom is found worldwide, although it is not as popular in the West as in other cultures. It can be found in abundance on fallen wood logs that are moist and porous, sometimes covered in moss. This fungi can be seen perched on wet mossy logs, listening to the songs of the forest. Usually a fleshy tan to dark brown, it has the structure of an ear (more or less) and is cupped in shape, sometimes even displaying vein-like structures.

HABITAT: It is usually found growing on living and dead elder trees but also found on a number of various other hardwoods all over the world. This mushroom tends to grow in more temperate climates and subtropical areas. In North America, it is mostly found from the east of United States down to the southern states.

DIFFICULTY LEVEL: Commonly seen and foraged.

TYPICAL SIZE: These small fungi are no more than 3 inches across—the size of a human ear or smaller.

NOTES: This mushroom is one of my favorites to incorporate into cooking, as it has a rich earthy taste and texture and is very nutrient dense. It can be foraged in abundance and cooked fresh without much preparation. It goes well with seafood and vegetarian or vegan Asian-style dishes. Tree ear mushrooms can be found in some Asian markets as well as farmers markets year round.

CONSERVATION STATUS: None.

LOOKALIKES: Auricularia cornea is in the same genus as the tree ear mushroom but is in fact a different species. It can grow a lot bigger than the typical tree ear and does not have the ear shape when fully matured. They can also be confused by their color and gelatinous texture, but can be distinguished by their skin thickness; this lookalike has a much tougher and thicker skin than the tree ear mushroom.

EDIBILITY AND USES: Commonly used as an edible in classic Chinese food dishes and all over the world in various cultures.

FUN FACT: This mushroom is shaped like an ear! It is also known as black wood ear or jelly ear fungus. Although its name may not be appetizing, the tree ear mushroom is a choice edible fungus that is popularly used in traditional Chinese medicine and in Asian dishes like hot and sour soups and clay pot meals. Tree ear mushrooms can be found worldwide, although they are not as easy to find in the West. Tree ear mushrooms can be bought in Asian markets as well as farmers markets year round.

TURKEY TAIL

COMMON NAME	SCIENTIFIC NAME
Turkey Tail	*Trametes versicolor*

ORDER: Polyporales

Found in abundance around the world, turkey tail is a well-loved polypore mushroom. I have found them many times on my forest walks and forages. There are many lookalikes and false turkey tails, but most I have found to be medicinal and safe to ingest after proper preparation. Turkey tail has a beautiful display of colors, especially when just fully matured, spanning out like a fan of turkey feathers from the stump of old trees. It carries the colors of autumn in its orange, yellow, and red bands.

HABITAT: Typically found in damp and temperate areas, and in abundance on decaying wood. Commonly found in the northeast to far east regions in the United States and Canada, as well as far west, along the Pacific Ocean. Turkey tail can also be spotted in some Canadian regions close to the United States border, such as Toronto, Victoria, and the province of Nova Scotia.

DIFFICULTY LEVEL: Very common.

TYPICAL SIZE: The cap can measure anywhere from 1 to 3 inches in width and 1 to 3 millimeters thick.

NOTES: This mushroom is now known for its amazing properties in fighting cancer, after noted mycologist Paul Stamets came out in 2019 with a study involving his mother using a pharmaceutical drug and high doses of turkey tail; he helped his mother's cancer go into remission within six months. This mushroom has been used in many cultures medicinally and has been made into tinctures and capsules by various commercial companies. (Unfortunately, many companies use the mycelial mass, which has a higher percentage of additives like brown rice flour, rather than the actual fruiting body.)

CONSERVATION STATUS: None.

LOOKALIKES: False turkey tail *Stereum ostrea* can take on the appearance of turkey tail but is larger in size and is a more crust-like fungus that grows on dead logs. Its brownish brick red coloring (which can become green over time with algae growth) along with its porous underside makes it visibly distinguishable from turkey tail.

EDIBILITY AND USES: This mushroom is inedible but is used widely for its cancer fighting and medicinal benefits.

FUN FACT: Turkey tail is now known for its amazing potential for aiding in the fight against cancer. Noted mycologist Paul Stamets came out with a TED Talk in 2019, speaking about how a combination of Western medicine and high doses of concentrated turkey tail helped his mother's cancer; within six months of the treatment, it went into remission.

This mushroom is just now being discovered and credited in the West, but it has been used medicinally in many cultures for centuries, as it can be made easily into tinctures, capsules, and powders.

VELVET FOOT

COMMON NAME	**SCIENTIFIC NAME**
Velvet Foot	*Flammulina velutipes*

ORDER: Agaricales

FAMILY: Physalacriaceae

This mushroom can go by many names but its most common is enoki. Wild specimens vs. cultivated species can look significantly different when foraging them. Some can even have a pink tint and be much bigger. Commonly seen in Asian dishes like udon and soups, this mushroom is commonly collected for its edible purposes. The caps can be tacky upon harvesting and have a flat round cap and short thick stem; when they are being cultivated, the CO_2 rich atmosphere makes them stretch for oxygen, and the stem can become long and thin with small caps rounding at the top.

HABITAT: Generally velvet foot likes to branch off from the stumps of ash, mulberry, and persimmon trees from September through March. This mushroom has been commonly spotted within northeast regions of the United States, including Chicago, Indianapolis, and New York. Canadian cities close to the eastern part of the United States, like Toronto and Montreal, can also have enoki. This mushroom has even been spotted in a few places in Mexico.

DIFFICULTY LEVEL: Moderately difficult to find.

TYPICAL SIZE: The caps can get up to almost 2 inches and the stem range from 1 to 3 inches tall.

NOTES: These mushrooms are heavily cultivated throughout the world and commonly found in Asian cuisines. They are also considered medicinal due to their antioxidant properties. Research has found that this mushroom suppresses cancer growth, and that the flamullin in particular inhibits tumors from growing.

CONSERVATION STATUS: None.

LOOKALIKES: Flammulina populicola is a common lookalike but has a lighter and tannish cap with a matching stem. It can also grow on hard old woods, unlike the velvet foot which prefers select trees like ash and mulberry.

EDIBILITY AND USES: Commonly used in Japanese cooking and Asian restaurants.

FUN FACT: The velvet foot mushroom is most commonly known as enoki. Note that there are wild specimens and cultivated species that can look significantly different than when foraged; this is because of the environment, CO_2, and oxygen. When cultivated and grown in jars, velvet foot mushrooms produce a gas that stifles them, making them stay small and stretch tall, reaching for the lid of the jar to get oxygen. They are then harvested in large clusters. In nature, velvet foot mushrooms can be quite big. Some may even have a pink tint in subtropical regions.

VELVETY EARTH TONGUE

COMMON NAME
Velvety Earth Tongue

SCIENTIFIC NAME
Trichoglossum hirsutum

ORDER: Geoglossales

FAMILY: Geoglossaceae

This mushroom is a long, tongue-like fungi that grows in grass, moss, and soil. It is black in appearance, and the stem, which can look wet or shiny around the base, is sticky to the touch. These mushrooms grow in single patches that can be scattered throughout one area. Some, although rare, can have relationships to hardwood trees, and can be found growing on them as well.

HABITAT: Typically found growing in grassy areas, soil rich fields, and on moss-covered hardwood or grounds.

DIFFICULTY LEVEL: Generally pretty rare to find.

TYPICAL SIZE: The cap is 0.3 inches long and the stem is generally 0.6 inches. The mushroom can get up to 2 inches in height.

NOTES: In certain parts of the United Kingdom this mushroom is used as ground control to show the quality of the grass in that area.

CONSERVATION STATUS: None.

LOOKALIKES: Geoglossum umbratile is a similar species and lookalike to the Glutinoglossum, which has a sticky cap but identical body; and the Trichoglossum, which has a velvety soft cap as well as an identical cap and stem.

EDIBILITY AND USES: Although not generally popular for culinary purposes because of its small size, this mushroom is edible and has been reported to be quite delicious.

FUN FACT: The velvety earth tongue is known for its black, sticky, and slightly hairy fruiting body. It protrudes straight up from the ground and grows a wavy flat tongue-like cap at the top. This unusual mushroom is also known as hairy earth tongue or black earth tongue.

WHITE CORAL

COMMON NAME
White Coral

SCIENTIFIC NAME
Clavulina cristata

ORDER: Cantharellales

FAMILY: Clavulinaceae

White coral is a common fungus that grows along the ground near conifers. It is bright white, and it forms branches that then split off into more branches, following a branching pattern similar to that of oyster mushrooms and actual coral. It is an edible species and has a rubbery texture that is not always pleasant. Other coral fungus species are present, but this is the only one that is white, making it a quick, easily distinguishable find when foraging.

HABITAT: Generally found on or near dead wood protruding from the ground, or moss-covered soil with or around other fungus throughout North America.

DIFFICULTY LEVEL: Generally common to find.

TYPICAL SIZE: This fungi is pretty small, growing up to 1 to 3 inches tall, with ½ inch branch-like extensions.

NOTES: This mushroom grows from the ground but can sometimes be seen growing on moist dead logs covered in moss. They can grow singly or in a large cluster or group.

CONSERVATION STATUS: None.

LOOKALIKES: The crown tipped coral mushroom looks almost identical to the white coral mushroom, but has shorter branches and a big round base. They do both happen to grow in similar areas around moss and fallen trees that have been covered with algae and moss.

EDIBILITY AND USES: None.

FUN FACT: White coral mushrooms are small in size and can appear to be somewhat of an alien in the mushroom world. They present no cap or stem; instead, their fruiting body branches out and then continues to branch again and again. It is an uncommon mutation, but it is this mushroom's natural structure. It is soft and gelatinous, and is even edible. It may have some cancer fighting medicinal benefits that should be studied further.

WINE CAP

COMMON NAME
Wine Cap

SCIENTIFIC NAME
Stropharia rugosoannulata

ORDER: Agaricales

FAMILY: Strophariaceae

This wine-colored mushroom is a heavy helper in the garden and can decompose pretty much anything! I love using this mushroom when working with compost and making some rich nutrient dense mulch for the garden. Wine cap mushroom carries a compound that can help break down the soil and create rich fertility in the flora. It is also considered a choice edible and can grow considerably tall and big, making it a lasting food worthy of cooking for various meals. The cap of this mushroom, as its name suggests, is a wine to burgundy red color. The stem is generally white to gray, with a thick ring around the middle of the base where the veil has broken.

HABITAT: Generally found in woody areas, open grassy fields, lawns, and around the edges of garden beds. Typically located on the east coast of the United States but can be found throughout North America.

DIFFICULTY LEVEL: Generally quite easy to find.

TYPICAL SIZE: The stem gets up to 7 inches tall, and the cap is 11 inches wide.

NOTES: This mushroom is commonly known for its edible use and ability to make rich, beautiful compost.

CONSERVATION STATUS: None.

LOOKALIKES: Some wine caps can be mistaken for agaricus or an agrocybe species of mushroom, yet those mushrooms almost always have a white cap and brown underbelly.

EDIBILITY AND USES: Commonly used as a choice edible in many countries, thanks to its large size and pleasant taste.

FUN FACT: The wine cap mushroom is known as the great composter! It is commonly found and propagated by cultivators and mushroom hobbyists for its amazing ability to turn backyard scraps into black gold.

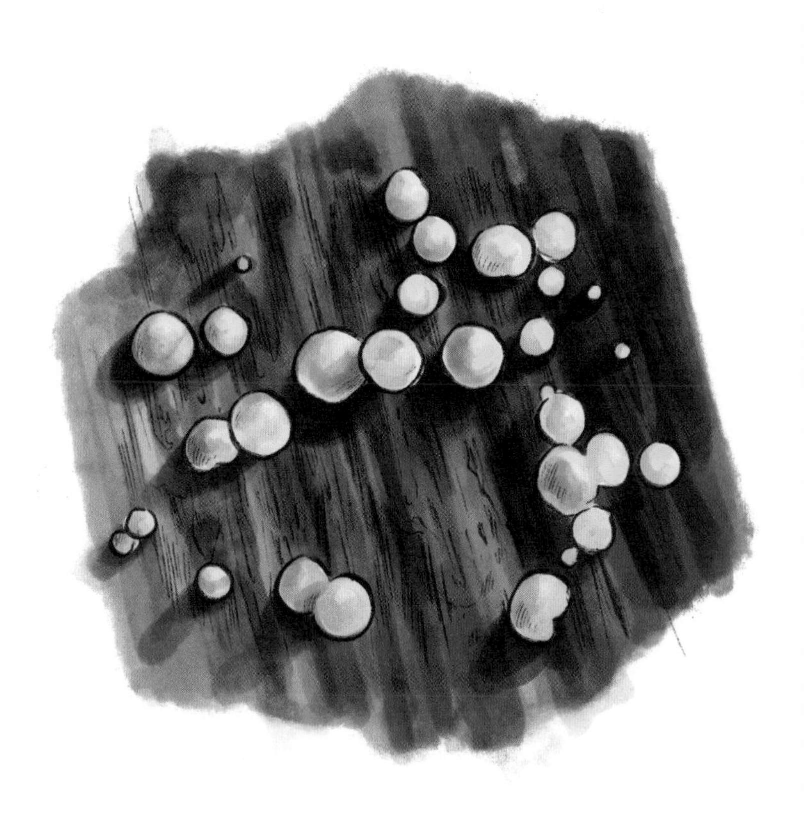

YELLOW FAIRY CUPS

COMMON NAME
Yellow Fairy Cups

SCIENTIFIC NAME
Bisporella citrina

ORDER: Helotiales

FAMILY: Helotiaceae

Yellow fairy cups are a cup-like species of fungus that grow on dead or decaying wood. They present a beautiful display of tiny gelatinous and porous bright yellow caps that divot, protruding down into the stump or base of the tree where they grow in small clusters and groups. Because of their small size they are not easily seen or collected, and are not edible.

HABITAT: Found in late summer and autumn, this species lives all over the world in various climates, and is found on decaying wood that has lost most, if not all, of its bark. Some have been spotted in Ontario, and westward to Alberta in Canada, while others have seen them from Washington down to California in the United States; they have even been spotted on the east coast.

DIFFICULTY LEVEL: Because of its small size it can be hard to find.

TYPICAL SIZE: This fungus grows to about 0.12 inches across and does not typically have a stem attached.

NOTES: Although this species has many lookalikes, the yellow fairy cups mushroom's distinctive color and size can help distinguish it from lookalikes. Other mushrooms may even have to be identified under a microscope or by looking at the shape of their spores.

CONSERVATION STATUS: None.

LOOKALIKES: Hymenoscyphus calyculus is a small yellow cup fungus that can be easily confused with the yellow fairy cups. It has the same color and small size, but unlike the yellow fairy cups, it has a small, skinny stem. It tends to grow in similar areas on tree bark and fallen logs.

EDIBILITY AND USES: None.

FUN FACT: These dainty little mushrooms are incredibly small. In fact, they can only grow up to 3 millimeters wide—about the width of a thumbnail. They emerge as clusters on decaying wood, making a pretty, colorful little scene.

GLOSSARY

ADAPTOGEN: A nontoxic substance, such as a mushroom or mushroom extract, which is said to increase the body's ability to resist the damaging effects of stress, and promote or restore normal physiological functioning.

AGAR: A malt medium that is used for the extraction of fungi tissue to grow on and be observed for mycology use.

ANAMORPHIC: Many mushrooms have sexual and asexual stages; an anamorphic stage is a mushroom that is starting an asexual phase.

BIOLOGICAL SPECIES: Biological species are species that can mate and reproduce to create offspring.

BRANCHING: Branching in reference to mushrooms or mycelium is the reaching of hyphae or other mushrooms towards a source of nutrients.

CAP: The cap is the top of the mushroom body containing the gills and spores, which are released through the underbelly.

COLONIZE: The process in which the mycelium network is growing.

CONK: A round or kidney-shaped polypore or shelf mushroom that can be hard or soft.

CONIFER: A type of evergreen tree or shrubs that produce cones.

CULTIVATE: A production of growing mushrooms or plants indoors for produce year round; a cultivation practice or cultivator would be one that produces mushrooms to sell.

DEAD WOOD: Rotting, decaying, or dead trees, logs, and stumps.

ENTHEOGEN: A psychoactive, hallucinogenic substance made of or derived from fungi or plants and used ceremonially.

FALSE: A term used to describe a lookalike or similar fungi/species mimicking the appearance of another mushroom.

FLUSH: When mature mushrooms grown in a group-like formation are fully fruited.

FUNGUS: Fungus is one of nature's kingdoms if it falls under being a eukaryotic organism.

FRUITING: When a mushroom has gone into its full mature phase and is starting to break its veil and release spores.

GILLS: The underbelly of the mushroom cap where fine lines of soft tissues that hold spores are kept from being released.

HYPHAE: Fine strands of mycelium that branch out to find new connections in the network.

LOOKALIKE: A false mushroom or one that is easily mistaken for another.

MEDICINAL MUSHROOMS: Mushrooms with adaptogenic or significant healing effects on the body. This includes species like reishi, shiitake, maitake, turkey tail, chaga, lion's mane, and more.

MYCELIUM: A thick white membrane and connective tissue that creates a web-like pattern providing sourcing of food, nutrients, and protection from certain pathogens while the mushroom is growing. The mycelial web is the strongest and one of the most vital parts of nature that helps create new connections with plants, animals, and even other humans that are in the area.

MYCOLOGIST: Someone who studies mushrooms and fungi.

MYCOLOGY: The study of mushrooms.

MYCORRHIZAL: A symbiotic and synergistic relationship between a fungus and a plant.

PARASITE (PARASITIC): An organism that lives on or in a host and receives its source of food from the host.

POLYPORE: A conk or solid belly mushroom that display large fruiting bodies on hard wood and decaying trees.

RHIZOMORPHIC: Mycelial growth invading or growing with the root structure of trees and tall plants.

STIPE (STEM, STALK): The base and structure of the mushroom which can take on a number of characteristics.

SPORES: Spores are shot out from the gills or underbelly of the mushroom and get taken by the wind; this is how they continue to spread their spores so they can survive as a species. Mushrooms are responsible for their own reproduction and do not need fertilization by another spore. Although this can be done by manipulation, which is how hybrids are created through cultivation or a natural reoccurrence.

SPORE PRINT: A cultivation technique used by mycologists, which involves placing the cap of a mushroom on a piece of aluminum foil and allowing the spores to settle onto it. The cap is then removed 24 hours later to reveal a print of the underbelly.

ABOUT THE AUTHOR AND ARTIST

Niko Summers is a fifth-generation herbalist from a long line of root workers and midwives. He became fascinated by fungi while studying their medicinal properties. In 2020, Niko founded Native Mushrooms, a mycology company that cultivates rare and medicinal species. When not studying herbs or fungi, Niko likes to go foraging in the woods or spend time in his garden. He is a Native San Franciscan, where he still lives with his dog. Find him on Instagram @nativemushrooms.

June Lee is a graphic designer and illustrator who enjoys everything communication design-related, from painting murals to branding design. Projects include the Truth Campaign for *The New York Times*, Pizzaz Pizza Festival, and Junk magazine. In her free time, June can be found reading with a cup of coffee in hand, or out hunting for the best coffee in the city. She lives in Brooklyn, New York. For more of her work, visit juneleedesign.com.

ACKNOWLEDGMENTS

I would like to thank my parents, my fiancé Joséphine, and my mentor and business partner James, who owns The Haight St. Shroom Shoppe, as well as John Whalen, Melissa Gerber, and Margaret Novak.

ABOUT CIDER MILL PRESS BOOK PUBLISHERS

Good ideas ripen with time. From seed to harvest, Cider Mill Press brings fine reading, information, and entertainment together between the covers of its creatively crafted books. Our Cider Mill bears fruit twice a year, publishing a new crop of titles each spring and fall.

"Where Good Books Are Ready for Press"

Visit us online at
cidermillpress.com
or write to us at
501 Nelson Place
Nashville, TN 37214